A JOURNAL OF
CONSCIOUSNESS AND TRANSFORMATION

ReVision

I0116428

CONTENTS

Integral Consciousness
Bahman Shirazi, Editor

1   Introduction
Bahman Shirazi

2   Integral Education for a Conscious Evolution
Anne Adams

7   Wholeness, Integration of Personality,
and Conscious Evolution in Integral Psychology
Bahman Shirazi

13   Atemporal Creativity: Evolution Beyond Lines & Spirals
Zayin Neumann

22   The Oneness (and One-ing) of the Way:
Using Both Hemispheres of the Global Mind
Sheri Ritchlin

28   Millennium Poem
Michael Sheffield

29   The Role of the Astrological Symbol System
in Understanding the Process of Evolutionary Growth
Armand M. Diaz

35   A Feminist Approach to Three Holy Women's Intuitive Processes:
Exploring Aspects of Transcendental Phenomenology
& Hermeneutics Through a Correlation of the Researchers' Experiences
Ana Perez-Chisti

47   Integralizing Goddess: A Philosophical & Practical Approach to Spiritual
Awakening Through the (R)evolutionary Five-fold Feminine Force
Chandra Alexandre

53   Haiku Sequence
Michael Sheffield

54   Transformative Travel: An Enjoyable Way to Foster Radical Change
Susan L. Ross

62   Unconditioned Mind (Poem)
Michael Sheffield

Cover image: Mandala by Sharon Till

Spring 2010 • Volume 32 • Number 1

What is ReVision?

For almost thirty years ReVision has explored the transformative and consciousness-changing dimensions of leading-edge thinking. Since its inception ReVision has been a vital forum, especially in the North American context, for the articulation of contemporary spirituality, transpersonal studies, and related new models in such fields as education, medicine, organization, social transformation, work, psychology, ecology, and gender. With a commitment to the future of humanity and the Earth, ReVision emphasizes the transformative dimensions of current and traditional thought and practice. ReVision advances inquiry and reflection especially focused on the fields presently identified as philosophy, religion, psychology, social theory, science, anthropology, education, frontier science, organizational transformation, and the arts. We seek to explore ancient ways of knowing as well as new models transdisciplinary, interdisciplinary, multicultural, dialogical, and socially engaged inquiry. It is our intention to bring such work to bear on what appear to be the fundamental issues of our time through a variety of written and artistic modalities. In the interests of renewal and fresh vision we strive to engage in conversation a diversity of perspectives and discourses which have often been kept separate, including those identified with terms such as Western and Eastern; indigenous and nonindigenous; Northern and Southern; feminine and masculine; intellectual, practical and spiritual; local and global; young and o

Artwork: Mariana Castro de A

Volume 32, No. 1 (ISBN 978-0-9819706-3-9)

ReVision (ISSN 0275-6935) is published as part of the *Society for the Study of Shamanism, Healing, and Transformation*.

Manuscript Submissions

We welcome manuscript submissions. Manuscript guidelines can be found on our webpage http://revisionpublishing.org.

POSTMASTER: Send address changes to
ReVision Publishing, P.O. Box 1855, Sebastopol, CA 95473.

Subscriptions

For subscriptions mail a check to above address or go to
www.revisionpublishing.org.

Individual Subscriptions

Subscription for one year: $36 online only,
$36 print only (international $72),
$48 print and on-line (international $84).

Subscription for two years: $60 online only,
$60 print only (international $96),
$79 print and online (international $115).

Subscription for three years: $72 online only,
$72 print only (international $108),
$96 print and online (international $132).

Institutional Subscriptions

$98 online only (international $134),
$134 print and online (international $191).

Please allow six weeks for delivery of first issue.

ReVision Abstracts

Vol. 32 No. 1   *Spring 2010*

Adams, A. (2010). Integral education for a conscious evolution. *ReVision*, *32*(1), 2-6. doi:10.4298/REVN.32.1.2-6

This paper provides a theoretical and experiential foundation for integral education and its contribution to the evolution of consciousness. Consciousness within an extended epistemology is viewed as manifesting and expanding through the development and integration of the physical, emotional, mental and spiritual intelligences. Examples from integral education programs are provided to illustrate how the curriculum and school environment purposefully creates a unique awareness of self, self in relationship with self and environment, self in relationship with others, e.g., community, family, nation and world, etc., as well as other life forms. An invitation is made to shift the focus of education from perpetuating our outdated cultural traditions to a reinventing of our human species.

Alexandre, C. (2010). Integralizing goddess: A philosophical and practical approach to spiritual awakening through the (r)evolutionary five-fold feminine force. *ReVision*, *32*(1), 47-52. doi:10.4298/REVN.32.1.47-52

Within the evolution of consciousness are developments that reflect awareness of an emergent movement inclusive of the rational and somatic, the goal-oriented and the meditative through new structures, forms and stages. In this article, the author adds to the conversation mythopoetic (rather than pre-rational) analysis and revelations of planetary consciousness inclusive of eco-feminist, paradoxical, and critical spirituality concepts, in order to afford us the opportunity for breakthroughs and new potentials toward sustainability on the journey. She does so by the grace of inspiration emerging at the crossroads of the integral and the retroprogressive known as the dark goddess.

Diaz, A. (2010). The role of the astrological symbol system in understanding the process of evolutionary growth. *ReVision*, *32*(1), 29-34. doi:10.4298/REVN.32.1.29-34

Astrology is too often assumed to be a relic of primitive ways of thinking, although as a discipline it has evolved over the centuries in a way that more or less reflects the cultural eras in which it has been rooted. This article suggests the use of C. G. Jung's concept of synchronicity and ideas from chaos theory as a means of updating our way of thinking about astrology and expanding it beyond the language of myth and determinism. Jenny Wade's model of the evolution of consciousness is explored to expand and refine astrology's scope through the problematic issue of astrological prediction. The potential value of this approach in therapeutic and evolutionary contexts is discussed.

Neumann, Z. (2010). Atemporal creativity: Evolution beyond lines and spirals. *ReVision*, *32*(1), 13-21. doi:10.4298/REVN.32.1.13-21

The work of Jean Gebser suggests that consciousness does not unfold linearly, but in leaps or mutations — what he terms spiritual evolution — and that the upcoming mutation will be atemporal and acausal, rather than linear or cyclical. For this deep intuition to be applied in any real way it requires a clearer elucidation of the process of spiritual evolution whereby novelty ingresses in actuality. This paper illustrates how Alfred North Whitehead's consideration of propositional feelings, intellectual feelings, and rational knowing we can find a coherent description of an originating impulse that he terms 'Creativity,' the process whereby leaps of novelty can occur. By combining the works of these two authors we can grow a deep and penetrating understanding of evolution and consciousness.

Perez-Chisti, A. (2010). A feminist approach to three holy women's intuitive processes: Exploring aspects transcendental phenomenology and hermeneutics through correlation of the researchers' experiences. *ReVision*, *32*(1), 35-46. doi:10.4298/REVN.32.1.35-46

Hidden in Holy Women's intuitive choices are the seeds for transformative potential that can uplift our lives. Women who benefit humanity can draw us closer to ourselves because their insight and action have sprouted multiple perspectives that lie dormant in our own experiences. This article presents the stories of Three Holy Women Processes using a feminist approach to transcendental phenomenology and hermeneutics. As a postmodern theorist, the author views her own experiences as a source of knowledge through a felt-sense body knowing arising out of the diverse intuitive impulses found in these Holy Women's lives. The research approach uses participatory synchronicity and correlation found in "lived experiences" which are brought into the light.

Ritchlin, S. (2010). The Oneness (and One-ing) of the World: Using both hemispheres of the global mind. *ReVision*, *32*(1), 22-27. doi:10.4298/REVN.32.1.22-27

When exploring the nature and meaning of integral consciousness, we are challenged to bring our emerging "global mind" to bear on questions calling upon an Eastern as well as a Western perspective. This article examines the integral consciousness of ancient Chinese sages as they appear in the third century BCE Ta Chuan, the "Great Treatise" on the I Ching, and its close relationship to an emerging worldview in the West. The text offers insights into how a deeply integrative consciousness can bring the power of human discovery and invention into balance with the wisdom and integrity of human conduct toward a personal and planetary flourishing.

Ross, S. L. (2010). Transformative travel: An enjoyable way to foster radical change. *ReVision*, *32*(1), 54-62. doi:10.4298/REVN.32.1.54-62

Sacred or secular, travel provides the sojourner with a process that is often enjoyable and has outcomes that rekindle the spirit, rejuvenate the body, and even ignite a shift in consciousness. Transformative travel involves consciously created and experienced travel that combines key elements to maximize the potential for transformation. This article introduces transformative travel as one way to foster a shift in consciousness. Transformation and transformative travel are defined and discussed, and conditions are delineated that can cause travel to be transformative. Six categories of transformative travel discussed: pilgrim, mystic, diaspora, initiate, secular, and learner.

Shirazi, B. (2010). Wholeness, integration of personality, and conscious evolution. *ReVision*, *32*(1), 7-12. doi:10.4298/REVN.32.1.7-12

This essay addresses the transformation and integration of personality and attainment of a fully integrated psyche devoid of unconsciousness in the context of integral yoga and psychology. First integral consciousness is defined and its relationship to the complementary processes of evolution and involution of consciousness are discussed. Then the notions of wholeness and integration of personality are discussed using parallels between selected approaches to depth psychology and aspects of Integral Psychology based on the Integral Yoga of Sri Aurobindo. It is argued that conscious evolution necessitates full integration of personality resulting in the eradication of unconscious psyche.

Introduction

Bahman Shirazi, Editor

This special issue of *ReVision* contains a selection of articles from the proceedings of the first Symposium on Integral Consciousness held in May 2009 at the California Institute of Integral Studies (CIIS) in San Francisco. CIIS was founded by Haridas Chaudhuri who was the first scholar to bring the teachings of the great Indian sage Sri Aurobindo on integral yoga and philosophy to the West.

The general themes of the symposium were 'evolution of consciousness and conscious evolution' as well as 'integrative health and healing'. Articles appearing in this issue are selections from several presentations on integral consciousness, integral education, Asian roots of integral consciousness, and the feminine divine.

In *Integral education for a Conscious Evolution*, Anne Adams, provides a theoretical and experiential foundation for integral education and its contribution to the evolution of consciousness. Examples from integral education programs are provided to illustrate how the curriculum and school environment can purposefully create a unique awareness of self in relationship the environment, community, family, nation and the world.

The article by the editor titled *Wholeness, Integration of personality, and Conscious Evolution in Integral Psychology* addresses transformation and

Bahman A.K. Shirazi, PhD., is a faculty member at the California Institute of Integral Studies in San Francisco and Institute of Transpersonal Psychology in Palo Alto, California.

integration of personality and attainment of a fully integrated psyche in the context of integral psychology and yoga drawing on parallels between selected approaches to depth psychology and aspects of integral psychology. It is argued that conscious evolution necessitates full integration of personality resulting in the eradication of the unconscious psyche.

The next article by Zayin Neumann titled *Atemporal Creativity: Evolution Beyond Lines and Spirals* discusses the work of Jean Gebser suggesting that consciousness does not unfold linearly, but in leaps or mutations and that the upcoming mutation will be atemporal and acausal, rather than linear or cyclical. The author illustrates how Alfred North Whitehead's consideration of propositional feelings, intellectual feelings, and rational knowing can provide a coherent description of an originating impulse termed 'Creativity', the process whereby leaps of novelty can occur.

Sheri Ritchlin's article *The Oneness (and One-ing) of the Way: Using Both Hemispheres of the Global Mind*, examines the integral consciousness of ancient Chinese sages as they appear in the third century BCE *Ta Chuan*, the "Great Treatise" on the *I Ching*, and its close relationship to an emerging worldview in the West.

The article titled *The Role of the Astrological Symbol System in Understanding the Process of Evolutionary Growth* by Armand Diaz explores the use of C. G. Jung's concept of synchronicity and ideas from chaos theory as a means of updating our way of

thinking about astrology and expanding it beyond the language of myth and determinism. Jenny Wade's model of the evolution of consciousness is explored to expand and refine astrology's scope through the problematic issue of astrological prediction.

In *A Feminist Approach to Three Holy Women's Intuitive Processes*, Ana Perez-Chisti presents the stories of three holy women's intuitive choices as seeds for transformative potential that can uplift our lives. Using a feminist approach to transcendental phenomenology and hermeneutics and as a postmodern theorist, the author views her own experiences as a source of knowledge through a felt-sense body knowing arising out of the diverse intuitive impulses found in these Holy Women's lives.

In her essay titled *Integralizing Goddess: a philosophical and practical approach to spiritual awakening through the (r)evolutionary five-fold feminine force* Chandra Alexandre discusses breakthroughs and new potentials toward sustainability and the mythopoetic analysis and revelations of planetary consciousness in the context of eco-feminist and critical spirituality.

Finally, in *Transformative Travel: An Enjoyable Way to Foster Radical Change*, Susan Ross introduces transformative travel as a way to foster a shift in consciousness. Transformation and transformative travel are defined and conditions are delineated that can cause travel to be transformative. Six categories of transformative travel are discussed: pilgrim, mystic, diaspora, initiate, secular, and learner.

Integral Education for a Conscious Evolution

Anne Adams

onsciousness is "an awareness of self, knowledge of one's own existence and that of the objective world ... a presence that is always there (Damasio, 2003, p. 184). This awareness of ourselves and our world, includes our thoughts, feelings, sensations, identity and worldviews (Schlitz, 2005, p. xl). For the purpose of powerfully connecting an integral education with the evolution of consciousness, consciousness is also viewed from an extended epistemology.

There are four theoretical models, which provide a way of relating to consciousness that supports the understanding of integral education and its contribution to the expansion of our awareness. Kegan (1994) delineates levels of thinking that represent qualities of consciousness. First level thinking has a single-point focus on what is immediately present;

Anne Adams, PhD, has designed and led seminars, workshops and educational programs for professional groups, individuals, corporations and educational institutions for more than 30 years. She has been a teacher, school director, university instructor, manager in an international educational corporation, and a business consultant to both large and small corporations. Anne has worked with many Fortune 50 companies, nationally and internationally. Her commitment is to have people create exciting, compelling futures and together bring those futures into reality. Phone: (650) 726-0231.

second level thinking focuses awareness of the self and one's own needs; third level thinking includes awareness of one's self in relation to others and fourth level thinking expands the

Integral education lays the foundation for an integral worldview by providing a consciously created cultural narrative which can powerfully help architect our 21st century world.

awareness to include a worldview that is systemic and complex, i.e., a consciousness that is aware of itself at the level of system (pp. 94-95).

Wilber (2000) has invented a powerful context for engaging with consciousness. His quadrant model includes the levels, i.e., matter, body, mind, soul, and spirit and the facets, intentional (interior-individual-subjective), behavioral (exterior-individual-objective), cultural

(interior-collective-intersubjective) and social, systemic (exterior-collective-interobjective). This framework highlights the integral mindset and its multi dimensional relationship with consciousness, including all aspects, the inside and outside, and the singular and plural.

Combs (1996) proposes three levels of organization of our experiential lives that further our understanding of consciousness. He has incorporated prior work done by Tart, Gebser, and Guenther in his levels (p. 257). The first level is states of mind, which contain feelings, emotions, moods, etc., the second is states of consciousness, which comprise experiential conditions such as dreams, ordinary awareness, meditation, etc, and the third is structures of consciousness, i.e. how the world is experienced and understood by human beings. The mental, physical and integral are examples of consciousness structures (pp. 257-264).

Beck (2002), through his Spiral Dynamics interpretation of the evolution of consciousness as eight spiraling and dynamic stages, points to the integral and holonic quality of the development of consciousness. Each phase represents 1) the prior 'living layer,' 2) the increased levels of complexity in both our external and internal worlds, and 3) the breakdown and reorganization that is inherent in life's dynamics. The spiral starts with stage 1) instinctive/ survivalistic values and moves through 2) magical/animistic, 3) impulsive/ egocentric, 4) purposeful/ authoritarian, 5) achievist/strategic, 6) communitarian / egalitarian, 7) integrative, and finally to 8) holistic (p.1). This explanation brings clarity to how consciousness has evolved throughout history inside of cultural, religious, social and economic influences. Claire Graves introduced the theoretical framework of Spiral Dynamics and offered this thinking about how to relate to the transformations in our consciousness.

The psychology of the mature human being is an unfolding, emergent, oscillating, spiraling process, marked by progressive subordination of older, lowerorder behavior systems to newer, higher-order systems as man's existential problems change (as cited in Beck, 2002, p.1).

How we educate has everything to do with how effectively individuals and cultures develop themselves in the face of life's dynamics, e.g. expanding their perception of themselves to include the deeply nuanced self that they are, and an ability to move their attention from themselves to others and their environment and ultimately to their world. Our education lays the foundation for our ability to confront and successfully relate to the breakdowns and reorganizations necessary to evolve our consciousness to subordinate lower-order behavior systems to newer, higher-order systems.

The premise of Integral Education presented here is consciousness develops and manifests through various domains of intelligence, uniquely as physical, emotional, mental and spiritual. Spiritual intelligence (SI) is often related to as the "creator" of consciousness (Combs, 1996), a context of connectedness - the individual with him or herself, the individual and the collective, the inner and outer, the silent and expres-

sive, the abstract and practical, etc. Our subjective experience of consciousness is represented through out mental intelligence while our objective experience of consciousness is revealed through our physical acuity (Chopra, 2005, p. 206). Emotional intelligence provides communication and relational channels, i.e., the conduit between subjective and objective experiences of consciousness. The development and integration of our

Art: Salma Arastu

mental, emotional, spiritual and physical intelligences brings greater understanding of our inner states of being, the vast network of communication throughout the body as well as expanding our experience of the outer world as well, i.e. our deep connections to one another, nature and all other life forms.

There is much correlation between integral education and its stated purpose with Kegan's levels of thinking, Wilber's quadrant model that integrates levels and facets of consciousness, Combs' states of mind, stages and structures of consciousness and Beck's Spiral Dynamics directed toward the integrative and holistic stages of human development..

Integral Education lays the foundation for an integral worldview by providing a consciously created cultural narrative which can powerfully help architect our 21st century world. A new quality of worldview literacy is being born, bring-

ing with it a transformed philosophical context. Our current global culture is no longer served by our historical ontological, epistemological, axiological and relational contexts (Laszlo, 2005; Senge, et al., 2004). The past approaches to education with their ensuing ways of thinking and being, values, beliefs, behaviors, relationships, etc., have created a world in which there are extensive examples of lower levels of both individual and collective consciousness. Examples of this are the state of the world economy, e.g. consumption, borrowing, and lending practices; unscrupulous business dealings, e.g. banking, real estate, corporate; media manipulation, a disregard for the earth's energy crisis and global warming; economic and human inequality, wars, prejudices and suppression of others. They illustrate Kegan's second level thinking focusing awareness on the self and one's own needs, minimal aspects of Wilber's Quadrants, the intentional (interior-individual-subjective) and behavioral (exterior-individual-objective) and the lowest levels of Beck's spiral - the instinctive/ survivalistic and impulsive/ egocentric stages of values consciousness. The integral worldview is not only a way of knowing and thinking, it is also a way of being, behaving, valuing and relating. The understanding and expression of the integral worldview promotes "a transdisciplinary perspective that emphasize[s] the intrinsic order and interdependence of the world in all its manifestations" (Banathy, 1996, p. 1).

Many scholars representing divergent viewpoints, ranging from science, business, medicine, education, spiritual communities, the arts, etc., have been advocating for some time that our collective cultural consciousness is insufficient for the world we are living in (e.g. Chopra, 2005; Laszlo, 2005; H. Smith, 2001; Maturana, 1999; Senge, et al.,2004; O'Sullivan (n.d); Pert, et al., 2005; Miller R., 2000, Mitchell, 2004; J. Miller,

2006; Ray, 1996; Hock, 1999; Wilber, 2003; Eisler, 2002; Morin, 2002).

Thomas Berry (1999), one of our recognized sages, was a historian and custodian of the wisdom of cultures and religions, from both the Eastern and Western traditions. At the same time, he had a depth of understanding of and appreciation for science and its ability to inform us from many different angles. His call to us at this time came from his deep commitment to the evolution of our collective consciousness. He declared,

We need to reinvent the human at the species level because the issues we are concerned with seem to be beyond the competence of our present cultural tradi-

he points to our fractured cosmology, i.e., "our loss of a coherent conception of ourselves, our universe, our relation to one another and our world" (p. 7). He envisions integral education as a way of shifting our consciousness to a planetary context. Clark (1997), another advocate of the integral approach to education, critiques the current established educational structure as not being equipped to cope with the speed and complexity of the major changes taking place in the world today. He, like O'Sullivan (n.d.), R. Miller (2000) and J. Miller (2006), also sees the systemic, integral perspective as needed to encompass and educate for the multiple purposes of education,

the foundational philosophy of integral education. The "complex is that which is woven together," (p. 34). What has been missing that is now being revealed is a paradigm that can truly embody the level of complexity that exists today– a complexity paradigm. This complexity paradigm creates an interdependent mesh that weaves together and binds unity and multiplicity (Morin, 2001), the interior and exterior, the individual and collective, and the cultural, social and systemic (Wilber, 2000). The manner in which paradigms initiate and take root is through individuals as they experience, sense, learn, know, think, converse and act. Paradigms are interiorized and culturally inscribed, most often through education (Morin, 2001).

An integral education provides 'master concepts' that inscribe a different quality of interiorized paradigm; one that reinterprets and expands the idea of "culture" by interconnecting unity and multiplicity, exterior and interior, individual and collective, social and cultural, and local and global. Integrality in education promotes identities that are whole at many levels of human expression from individual to planetary. Attention on the ontological quality of education, expressed in the development and integration of the physical, emotional, mental and spiritual intelligences, provides a foundational interwoven and resilient 'network' of intelligences – a wellspring for consciousness.

The integral worldview is not only a way of knowing and thinking, it is also a way of being, behaving, valuing and relating.

tions either individually or collectively. What is needed is something beyond existing traditions to bring us back to the most fundamental aspect of the human: giving shape to ourselves (as cited in Swimme & Berry, 2005, p. 578).

Berry, from the wisdom of his octogenarian perspective, saw the necessity of societal reinvention, not only because of the lack in our existing traditions, but also because of the inseparability of who we are with the quality of future we are capable of bringing forth.

> Our own future is inseparable from the future of the larger community that brought us into being and sustains us in every expression of our human quality of life...emotional, aesthetic, intellectual, sense of divine, as well as in our physical nourishment (as cited in Swimme & Berry, 2005, p. 580).

Berry is joined by others whose voices give strong resonance and resolve to their commitment to a transformation of consciousness. Jean Houston (2000), an integral scholar; psychologist, philosopher and spiritual activist, says we can no longer wait to reinvent a story to equip us to live in a world that today is no longer served by former ways of knowing and being. O'Sullivan (n.d.), professor emeritus of transformative learning, represents that perspective as

within individual, relational, communal, global and planetary contexts.

Morin (2001, 2002), another esteemed elder voice of philosophical wisdom, has contributed his knowledge of complexity, culture and paradigm creation to the reconstructing of education for the future. He understood that our current paradigms of "fragmentation, disjunction, separateness, which are reflected multi-dimensionally, make it impossible to grasp that which is woven together" (p. 38). We require a "paradigmatic change in the way we organize knowledge" (p. 29), "we need a paradigm compatible with complex knowledge to crystallize and take root," (p. 28). He distinguished a paradigm as the promotion and selection of master concepts of intelligibility to be integrated into a socio-cultural discourse which can alter the collective awareness.

Integral education offers a new paradigm for education that serves the depth of complexity in a world whose boundaries extend far beyond its parameters. We are in a world that is, as Morin defined it, "complex, multidimensional, planetary, global, transnational and polydisciplinary" (p. 29) and requires an extensive shift in consciousness in order to be sustainable. Morin's views in the domain of complex thought parallel much of

McCraty, Atkinson & Tomasino (2001) have found that

> Consciousness is impacted by the degree of mental and emotional coherence experienced. When they are out-of-phase, overall awareness is reduced. Conversely, when they are in-phase, awareness is expanded. This interaction affects us on a number of levels: Vision, listening abilities, reaction times, mental clarity, feeling states and sensitivities are all influenced by the mind and emotions integrating and coordinating (pp. 51-52).

An integral education is purposefully designed to focus attention on the students' awareness of themselves, through an approach that attends to the development of the physical, emotional, mental and spiritual domains. This self-awareness, through an integrated education, self reflective exercises, practices and conversations, naturally grows an ability to consciously attend to the surrounding world. An example of the impact these programs have had on their students, their community and the way they think and act toward one another, came from an observation a parent made about how an awareness accessed through the body shifts the consciousness of the student *and* the community in which the education is taking place.

She remarked after observing a class in Awareness through the Body,

> I saw that they love to move and be together by moving their bodies and experimenting [with] things. They [would] see that actually their body is the same as their friend's body and there is no difference and that it is the same matter. This is very interesting to see that they understand this. They could be able to respect the others as themselves.

She followed with a second observation, relating the expanding awareness of the students' and their relationship with their own bodies to the resulting consciousness in the community.

> I have to add that around here, there was never an incidence of violence, never mistakes or hurting each other. I never have experienced that in all these 13 years we are here. There is even no talk about it that they would like to hurt someone. They were able to respect the others as themselves.

Linda Olds (1992), as a psychologist and system theorist, validates this observation as she relates to the "body as a context for knowing."

> Our knowledge from its onset is also embodied, embedded in our kinesthetic relationship with reality and in the connection of our bodies to the physical world. Our bodily based experience of moving and interacting in the world impacts our ability to understand our world as much as our abstract intellectual thinking (p. 8).

Another integral program provides weekly opportunities for young people to come together to learn and practice silence, contemplation and meaningful, connecting communication with their peers. These sessions educate students in a respect for life and human beings and learning to hold in high regard the experiences and expressions of others. Times vary from 15 minutes for elementary, 30 minutes for middle to 1 hour for high school students.

One student offers this personal account of her experience,

> It is one of the few chances that we have to escape the hectic pace of our world and to think about things that matter. It does not take long for superficial thoughts to yield to deeper ones. It is a time to think about problems that we are facing personally, or those that the world is being confronted with. It is a time to think about what we believe in, and what we stand for. With none of the usual distractions influencing us, the meetings are a time when we can truly think for ourselves. We are not only taught to think for ourselves, but we are also given the power and responsibility to follow our hearts. (Note from Student Council Meeting, 2003)

A parent adds her perspective,

> Our children are so inundated with graphics images, extracurricular activities, please to "hurry up," "do this" and "stop that." What a refuge these meetings together offer them. It offers a chance to shed outside pressures and listen to their inner voices, to commune with their peers and teachers during times of celebration and sorrow, to reflect and to dream. The result I believe produces individuals who are mature, insightful and inquisitive. (Note from Parent Council Meeting, 2004)

An integral education is purposefully designed to focus attention on the students' awareness of themselves, through an approach that attends to the development of the physical, emotional, mental and spiritual domains.

These practices create a self knowledge and trust from an early age. The peer groups engaged in this quality of silence, listening and speaking provide human beings with direct experiences of their relatedness and connection with themselves and other people. Powerful, authentic listening gives rise to powerful authentic speaking; powerful listening and speaking give rise to a more powerful conscious engagement in life.

Parents of students in integral programs comment,

> The teachers are educated to relate to the children through their higher self. They work hard to see the children in that spiritual place ... allowing children to become themselves ... giving them room to grow up and find themselves rather than trying to be someone they are not.

The attention paid to the development of emotional intelligence (EI) in the integral education curriculum is extensive. The schools work with students to learn to handle upsets among themselves, develop meaningful relationships among peers and teachers, to care for one another and instill a powerful sense of who one is in their own right. An environment of respect, trust, authenticity and acceptance is the reality created.

The physical development of the body is powerfully impacted by experiences like these, e.g., awareness through the body, being able to be both silent and reflective and self expressive with one's teachers and peers. These integral educational experiences require 'an open exchange between the organism and its environment' (Lipton, 2008, p.16), 'living in safety and love effects the body and mind' (Lipton, 2008, p.13), and ultimately the quality of awareness of both individuals and communities. Integral education literally supports the growth of a different quality of physical structure.

Consciousness expands through purposeful everyday practices and conversations that sharpen awareness of self and others. Conversations create our reality. Reality, inside of an integral education approach, takes on an inclusive nature, a both/and quality as opposed to the either/or dualism, e.g. separateness, fragmentation, that has been engrained in our current educational reality, that we so often take for granted as 'the way it is.' Stephanie Pace Marshall (2005), an internationally known educator, shares her experiences in this discourse.

> The nature and quality of our children's minds will shape who they become, and who they become will shape our world. Unfortunately, the world now being mapped into the minds our children is one of scarcity, fragmentation, competition, and winning. Our current story conceives learning as a mechanistic, prescribed, and easily measured commodity ... This narrative could not be more wrong. (p.12)

Engaging in conversation inside of an integral educational context supports a conscious evolution. "There is a different view of the world - a different view of the self and of others–a different worldview . . . as consciousness evolves" (Wilber, 2000, p. 132).

Our world today requires a different kind of human being: one who can think, create, imagine and act, with flexibility, adaptability and resiliency, in an extremely complex world; one whose spirit is vital and engaged and whose body is vibrant and healthy; one who can feel deeply and 'be present' to life, i.e., be aware in the moment and know how to move with and coordinate action in a highly diverse and accelerated world (Adams, 2006). Educating within an integral framework provides the foundation to develop and integrate the mental, physical, emotional and spiritual intelligences in order to go beyond our existing educational traditions to bring us back to the most fundamental aspect: giving shape to ourselves – reinventing ourselves at the species level (as cited in Swimme & Berry, 2005).

The purpose of this article is to invite readers to acknowledge the current levels of consciousness being created in our communities, states, nations and world through education as it exists today. It is time to tell the truth about how much our current educational philosophy – rather than expanding our consciousness, e.g. opening up our ability to self reflect, to see and discern our world – keeps us myopic and unconscious, hiding behind concepts, presuppositions, interpretations and justifications learned in our schools.

We have a choice to make. Is it the transformation of our relationship with and approach to education or a fast paced world filled with chaos, complexity and possibility that does not have a sufficient level of collective consciousness to engage intelligently with it?

References

Adams, A. (2006). *Education: From conception to graduation - a systemic integral approach.* www.cop.com/wisdompg.html

Banathy, B.H. (1996). Systems inquiry and its application in education. In D.H. Jonassen (Ed.), *Handbook of research for educational communications and technology.* New York: Simon & Schuster.

Beck, D. (2002, Fall/Winter). The never-ending upward quest. [Electronic version] *What is Enlightenment*, 22. Retrieved November 3, 2005 from www.wie.org/j22/beck.asp

Berry, T. (1999). *The great work: Our way into the future.* New York: Crown.

Chopra, D. (2005). Timeless mind, ageless body. In Consciousness and Healing, M. Schlitz, T. Amorok, & M. S. Micozzi (Eds.), *Consciousness and healing: Integral approaches to mind-body medicine* (pp. 201-211). St. Louis, MO: Elsevier.

Clark, E. T., Jr. (1997). *Designing and implementing an integrated curriculum: A student-centered approach.* Brandon, VT: Holistic Education Press.

Combs, A. (1996). *The radiance of being: Complexity, chaos and the evolution of consciousness.* St. Paul, MN: Paragon House.

Damasio, A. (2003). *Looking for Spinoza: Joy, sorrow, and the feeling brain.* New York: Harcourt.

Eisler, R. (2002). *The power of partnership.* Novato, CA: New World Library.

Hock, D. (1999). *The birth of the chaordic age.* San Francisco: Berrett-Koehler.

Houston, J. (2000). *Jump time: Shaping your future in a world of radical change.* New York: Jeremy P. Tarcher.

Kegan, R. (1994). *In over our heads: The mental demands of modern life.* Cambridge, MA: Harvard University.

Laszlo, E. (2005, July). Interesting times: Paradigm-shift in science; bifurcation in society. Paper presented at The Consciousness & Healing Conference, Institute of Noetic Sciences, Washington, D.C.

Lipton, B. (2005). *Biology of belief: Unleashing the power of consciousness.* San Rafael, CA: Mountain of Love/Elite Books.

Marshall, S. P. (2005). A decidedly different mind. Shift: At the Frontiers of Consciousness, 8, 10-15.

Maturana, H. R. (1999, Winter). The biology of business: Love expands intelligence. In Reflections, the SOL Journal on Knowledge, Learning and Change. Volume 1, Number 2, Cambridge, MA: MIT Press.

McCraty, R., Atkinson, M., & Tomasino, D. (2001). *HeartMath research.* HeartMath Publication, No. 01-001. Boulder Creek, CA. Retrieved July 3, 2005, from http://www.heartmath.org/research/research-publications.html

Miller, E. K., & Cohen, J. D. (2001). An integrative theory of prefrontal cortex function. *Annual Review of Neuroscience*, 24, 167-202.

Miller, J. (2006). *Educating for wisdom and compassion: Creating conditions for timeless learning.* Thousand Oaks, CA: Corwin Press.

Miller, R. (2000). *Caring for new life: Essays on holistic education.* Brandon, VT: Holistic Education Press.

Mitchell, E. (2004, February). Creating a wisdom society. Paper presented at the Institute of Noetic Sciences members meeting, Petaluma, CA.

Morin, E. (2001). *Seven complex lessons in education for the future.* Paris: UNESCO.

Morin, E. (2002). Introduction to complex thought. Unpublished Manuscript.

Olds, L. (1992). *Metaphors of interrelatedness.* Albany, NY: SUNY Press.

O'Sullivan E. (n.d.). Integral education: A vision of transformative learning in a planetary context. Retrieved May 2, 2005, from http://tlc.oise.utoronto.ca/insights/integraleducation.html.

Pert, C. B., Dreher, H. E., & Ruff, M.R. (2005). The psychosomatic network: Foundations of mind-body medicine. In M. Schlitz, M. T. Amorok, & M. S. Micozzi, (Eds.), *Consciousness & Healing* (pp. 61-78). St. Louis, MO: Elsevier.

Ray, P. (1996, Spring). The rise of the integral culture. *Noetic Sciences Review*, 37, 4.

Schlitz, M., Amorok, T., & Micozzi, M.S. (2005). *Consciousness and healing: Integral approaches to mind-body medicine.* St. Louis, MO: Elsevier.

Senge, P., Scharmer, C.O., Jaworski, J., & Flowers, M. (2004). *Presence: Human purpose and the field of the future.* Cambridge, MA: Society for Organizational Learning (SOL).

Smith, H. (2001). *Why religion matters: The fate of the human spirit in an age of disbelief.* San Francisco: Harper.

Swimme, B., & Berry, T. (2005). The ecozoic era. In M. Schlitz, T. Amorok, & M. S. Micozz, (Eds.), *Consciousness and Healing* (pp. 513-529). St. Louis, MO: Elsevier.

Wilber, K. (2000). *A brief history of everything.* Boston: Shambhala.

Wilber, K. (2003). Integral. Retrieved September 30, 2003 from http://www.integralnaked.org/integral.shtml.

Wholeness, Integration of Personality, and Conscious Evolution in Integral Psychology

Bahman Shirazi

I n integral yoga, philosophy and psychology, consciousness is understood as the essential underlying nature and structure of reality (Chaudhuri, 1977, pp. 33-42). In integral psychology consciousness is defined as the basic structure of the psychic microcosm, as well as the cosmic macrocosm. According to Chaudhuri (1975), "the psyche and the cosmic whole are inseparably interrelated. The universe is in ultimate analysis the psychocosmic continuum" (p. 233).

According to Sri Aurobindo (1970),

> consciousness is a reality inherent in existence. It is there even when it's not active on the surface, but silent and immobile; it is there even when it is invisible on the surface, not reacting on outward things or sensible to them... Even when it seems to us to be quite absent and the being to our view unconscious and inanimate (p. 234).

While Being is the ultimate ground of existence, consciousness is manifested on physical, vital, and mental planes in embodied forms. For Sri Aurobindo (1997) consciousness is not synony-

mous with mentality but indicates a self aware force of existence of which mentality is a middle term; below mentality it sinks into vital and material movements which are for us subconscient; above, it rises into the supramental which is for us the superconscient. But in all it is one and the same thing organising itself differently (p. 88).

The human being is then an embodi-

as Being (esse in Latin) or ground of existence (existere) is beyond phenomenal reality of time/space/matter and sentient existence (being) and ordinary consciousness. According to Meher Baba (1997) there are further subtleties in the 'beyond world' itself. He refers to Sufi and Hindu terms (wara' al-wara' and parat-para-Brahman respectively) that point to what lies even beyond the 'beyond' where there is not only no space, time or matter, but no consciousness as such.

Because of this absence of conscious-

tion of the collective unconscious, by its conscious counterpart, the Supermind.

Involution of Consciousness

It is generally held among a number of philosophical and spiritual traditions that there are two basic orders of reality: the transcendent, or ultimate reality, and the immanent, or phenomenal reality. The transcendent realm, also known

The ultimate aim of integral yoga is to eradicate the unconscious dimension of the human psyche and thus achieve a fully integrated conscious psyche.

ment of various spheres of consciousness ranging in density from the densest to potentially most luminous strata. The ultimate aim of integral yoga is to eradicate the unconscious dimension of the human psyche and thus achieve a fully integrated conscious psyche. This implies a full integration of not only the personal unconscious —the 'shadow'— into the 'ego' or the conscious part of personality as is proposed in Jungian psychology as part of the individuation process, but also the complete eradica-

Bahman A.K. Shirazi, PhD., is a faculty member at the California Institute of Integral Studies in San Francisco and Institute of Transpersonal Psychology in Palo Alto, California. Bahman is interested is in the processes of psychospiritual development and the interface between psychology and spirituality in general, and integral yoga and psychology in particular. Contact: bshirazi@ciis.edu. 1453 Mission St., San Francisco, CA 94103, (415)575-6253; (415)648-0550

ness, Being is not aware of itself, i.e. has no self-consciousness. This seed of un-knowing thus gives rise to a Divine desire for, and will to, self-knowledge which marks the beginning of the process of creation of the phenomenal world resulting in involution and evolution of consciousness. In Indian spiritual teachings three components or qualities comprise the 'beyond' state of transcendent reality: Sat (Being), Chit (Consciousness) and Ananda (Joy). Being (sat), or the ground of existence, is in a sense devoid of self-consciousness. While this ground of existence may be said to be devoid of dynamic self-consciousness, it has involved in it the seed of the highest possibilities of consciousness. Through the processes of the evolution of life on Earth this seed unfolds progressively to manifest increasing levels of consciousness through inorganic matter, to organic/vital, and then to mental consciousness. In the human being, self-consciousness fully manifests, and along with it, so does the potential for self-knowledge and self-perfection.

When this transcendent non-dual reality becomes manifested, the immediate result is a magnanimous cosmic polar structure one aspect of which Sri Aurobindo called the Supermind (called 'absolute consciousness' by Meher Baba), and the other side the Inconscient ('absolute unconscious'). This polar structure then becomes the basis for a multidimensional dialectical energetic exchange known as the processes of involution and evolution.

Involution is the descent, or outpouring and emanation of light, energy, and consciousness into ever-expanding spheres which, over the course of the process, diminish in intensity and luminosity resulting in the densest layers of matter at the outermost regions where light no longer penetrates. The involutionary process progresses in three successive stages. The first order is known as the Mind (not to be confused with thinking or other mental processes resulting from brain activity). It is this 'Mind' which in the Buddhist bible Dhammapada is referred to as the forerunner of all phenomena. Because Mind is antecedent to all other orders of reality, it is also referred to as the Causal realm in some esoteric traditions such as theosophy. Sri Aurobindo has distinguished between several orders of the Mind: the Supermind, the Overmind, the Intuitive Mind, the Illumined Mind and the Higher Mind in successive orders of involution down to the ordinary mental plane which humans have access to.

The next order of involution creates what is known as the subtle-physical or 'Vital' sphere. Also known as the 'Astral' world, the vital planes are the planes of manifestation of energy in their subtlest forms. The higher reaches of the

Biological evolution is only a part of the larger arc of evolution of the psyche on the physical plane.

vital planes are in contact with the lowest of mental planes, and the lowest vital planes are tangent to the highest strata of the gross-physical plane. The gross physical planes are generally classified as gases, liquids and solids, in order of increasing density. Superheated or highly energized gases result in 'ether' which borders on the vital planes.

As microcosmic beings, humans are capable of experiencing all of these planes by perception through corresponding senses. According to Meher Baba in the Mental planes only one sense exists—vision. In the Vital or subtle-physical planes two more senses become available, hearing and smell, and in the gross-physical plane the final two, taste and touch can be experienced. Thus in what Sri Aurobindo called 'outer-being' or surface consciousness we are capable of experiencing the world and constructing our reality based on the interaction of the five senses with the mind (also known as the knowing sense in Buddhism). In the 'inner-being' taste and touch which correspond to the gross-physical world cease to exist, but the other three senses (visual, auditory and olfaction) are extended (also known as celestial senses) and can be experienced through clairvoyance.

Evolution of Consciousness

Biological evolution is only a part of the larger arc of evolution of the psyche on the physical plane. In the Hindu tradition Brahman refers to the transcendent universal principle which is impersonal. Atman, on the other hand is the transcendent principle of individuality. In phenomenal actuality Atman does not 'exist', it is transcendent to time and space (thus the principle of Anatman or no-Self in Buddhism). As the cosmic consciousness becomes polarized to give rise to the process of creation from the Ground of existence, so does the seed of individuality (atma) or the human soul begins a 'separation' and the subsequent dialectical process with the ultimate soul of the universe (the Oversoul) which sets it on a journey of descent into creation and then a return journey or ascent to the original ground, or the Oversoul, in a journey of ever-increasing consciousness. Perhaps a more appropriate analogy is an outward journey from the cosmic core to matter and back inward through the process of evolution.

Just as the ultimate reality contains the seed of inconscience, so does matter (inconscient) carry in its innermost core the seed of the superconscient. The Chinese symbol for yin-yang is a depiction of this polarity/potentiality principle. This superconscient principle is at the core of matter and responsible for the upward evolutionary processes of the soul. While the masculine principle Purusha refers to the cosmic transcendent intelligent consciousness, the term Prakriti refers to the principle behind the evolutionary processes. Prakriti is the Feminine Divine intelligence-force responsible for all of the evolutionary processes.

In the course of creation the human soul is polarized into two principles or aspects. One aspect is situated at the core of matter and evolves up through the successive stages of evolution through material, vital, and mental planes. The other descends down from above, so to speak, to form the Jivatman or the energetic aspect. One might visualize this interplay by imagining an oval ring the two ends of which are the Jivatman at the top and the evolving soul (called the Psychic Being by Sri Aurobindo) at the core with (Kundalini) energy ascending from one to the other and back. At the culmination of the journey the two aspects become unified.

The intelligence inherent in matter operates through various instincts in all stages of evolution from mineral to organic, and from life to mind. As a soul journeys through the course of evolution and accrues experience, it becomes increasingly conscious. At the human level with the emergence of the upper cortex self-consciousness takes a mental turn. However, the ordinary human being at the mental stage of evolution is still highly instinctual albeit mentally self-aware. Sri Aurobindo's phrases "man is a transitional being" at "the crest of the evolutionary wave" emphasize that evolution has by no means reached its zenith or ultimate goal. Now human beings are in a position to consciously evolve and accelerate the processes of evolution into embodiment of higher reaches of the mind resulting in vaster consciousness up to the point of eradication of the unconscious portions of the psyche and attaining integrality and super-consciousness.

Wholeness and Integration of Personality

The interplay between consciousness and unconsciousness is at the core of the phenomenal world, and the reconciliation of these is a common goal of certain Western schools of depth psychology, notably Freud's psychoanalysis, Jung's analytical psychology, and Assagioli's psychosynthesis. The discovery of the 'unconscious' is perhaps the greatest achievement of the late 19th Century psychology. Later developments in depth psychology never disputed the existence of the unconscious, but further refined and extended it. The early transpersonal psychologists Carl Jung and Roberto Assagioli extended the range of Freud's unconscious to the collective dimension (Jung) and to the higher-unconscious (Assagioli), further highlighting the narrowness of the range of ordinary waking consciousness and the vastness of what lies beyond the conscious dimensions. Both Jung and Assagioli aimed at achieving wholeness of personality through integration of the

conscious and the unconscious dimensions of the psyche.

Freud made extensive use of the emerging concept of the 'Unconscious' in late 19th Century Europe in his formulations of the human psychological makeup. Jung extended Freud's idea of the individual unconscious, a repository of the psychologically repressed content, by introducing the idea of the collective unconscious which holds together groups and ultimately the entire human race. Jung was interested in maintaining a healthy balance between the unconscious and conscious aspects of personality. His process of individuation can be summarized as a harmonious reconciliation between the ego, or the conscious aspect of personality, and Self, the central archetype of the psyche which by definition belongs in the unconscious realm.

Jung (1959) held that an imbalance either in the form of ego inflation, or the domination of the ego by the

this reasoning, vast amounts of psychic energy would be required to overcome the unconscious completely. It was the goal of Sri Aurobindo's yoga to tap into such vast reservoir of energy, the highest possible Consciousness-Force, he called the Supermind.

Roberto Assagioli, the Italian psychiatrist who was an early associate of Freud, is not as well known as Freud or C.G. Jung. However, his framework called psychosynthesis, which combined empirical, depth, humanistic and transpersonal psychologies at once is, in this author's view, the most comprehensive Western psychological and psychotherapeutic system which in outline is strikingly close to the psychological dimension of Sri Aurobindo's integral yoga—integral psychology. Assagioli's comprehensive model of human personality is complemented with a rich array of practical techniques and processes for growth, development and integration of personality.

Sri Aurobindo's phrases 'man is a transitional being' at 'the crest of the evolutionary wave' emphasize that evolution has by no means reached its zenith or ultimate goal.

Self is equally undesirable. For him, it was important that the conscious and unconscious parts of the psyche stay in a dynamic balance. Inflation occurs when the ego oversteps its boundaries; at the same time, over-dominance of the Self was deemed a 'psychic catastrophe' resulting in poor psychological health. Jung did believe that it is important that the personal unconscious, or the shadow, be integrated into consciousness as much as possible, but he did not believe it was possible to eradicate the unconscious part of the psyche completely. Jung has been criticized for (at least at some point in his life) believing that due to limited amount of psychic energy (libido) for each individual, expansion of consciousness would have resulted in a 'vast but dim' state of consciousness. The dimness, therefore, would imply that the total eradication of the unconscious was not possible. Based on

Assagioli who incorporated in his model some of the best features of Freud's and Jung's contributions, introduced the idea of the 'higher unconscious' (absent in Freud's model and only implicit in Jung's framework), and called its organizing principle the Higher Self. He also introduced the idea of the 'middle unconscious' in his famous 'oval' diagram depicting his overall structure of human personality (Assagioli, 1971). While Assagioli's concept of the 'lower unconscious' is the same as Freud's concept of the unconscious (and Jung's personal unconscious or shadow) as the storehouse of dynamically repressed materials, his middle unconscious was added to account for what is not in the immediate conscious awareness, yet not dynamically repressed and available for recollection at will without any resistance or defense mechanisms. Assagioli's 'higher unconscious' referred explicitly to the human spiritual potential which could be made conscious and integrated into conscious personality just as the 'lower unconscious' would need to be made conscious and integrated to achieve complete integration and wholeness of human personality. The Higher or 'Transpersonal Self' would be crucial as a catalyst to make this integration possible.

Beginning in the mid 1920s Assagioli developed ingenious insights into the nature of the relationship between psychological and spiritual development and pointed out a number of psychological issues arising before, during and after spiritual awakening. However, his two major works were not published until the mid 1960s and later. Assagioli's first book titled Psychosynthesis (Assagioli, 1971/c1965) outlines his basic model of human personality as well as numerous techniques for catharsis, critical analysis, self-identification, dis-identification, development for the will, training and use of imagination, visualization and many more, all as part of the psychosynthesis work aimed at the integration of personality. His other important work, The Act of Will (Assagioli, 1973), outlines a number of types of will such as the strong will, the skillful will, the good will, the transpersonal will and the universal will, and how they can be utilized for various purposes. This is an important contribution to Western psychology and psychotherapy as other Western psychologists have rarely covered the topic of will which is the key to practical application in healing and psychospiritual development.

Although the idea of the evolution of consciousness does not play an explicit role in Assagioli's model, he has created a comprehensive system of depth psychology synthesizing the major contributions of both Jung and Freud. He also introduced the idea of multiple sub-personalities and believed that the soul or the Higher Self has a cardinal role in harmonization and integration of various aspects of personality. Assagioli developed the closest parallel system in Western psychology to that of Sri Aurobindo's integral yoga psychology.

Integral Yoga Psychology, Conscious Evolution, and Integration of Personality

Throughout our lives we carry in us a tremendous amount of unconscious energy. In fact, before birth we are mostly unconscious and still undifferentiated

With sustained spiritual practice we become 'awakened' and thus are able to transform the cycles of unconsciousness, attain continuous consciousness and ultimately overcome the cycles of death and rebirth.

from our mother. After being born a typical human infant spends the majority of his/her time in sleep. Eventually, the amount of sleep per 24 hours decreases and we manage to spend about two-thirds of our time in waking consciousness and the rest in various stages of sleep from dreaming to deep sleep. Finally, we die back into relative unconsciousness. It is believed that with sustained spiritual practice we become 'awakened' and thus are able to transform the cycles of unconsciousness, attain continuous consciousness and ultimately overcome the cycles of death and rebirth.

According to Sri Aurobindo (1992):

In the right view both of life and of Yoga all life is either consciously or subconsciously a Yoga. For we mean by this term a methodised effort towards self-perfection by the expression of the secret potentialities latent in the being and — highest condition of victory in that effort — a union of the human individual with the universal and transcendent Existence we see partially expressed in man and in the Cosmos. But all life, when we look

behind its appearances, is a vast Yoga of Nature who attempts in the conscious and the subconscious to realise her perfection in an ever-increasing expression of her yet unrealised potentialities and to unite herself with her own divine reality. In man, her thinker, she for the first time upon this Earth devises self-conscious means and willed arrangements of activity by which this great purpose may be more swiftly and puissantly attained (p.2).

In a paper titled: 'The Unconscious in Sri Aurobindo,' Indra Sen (n.d.) who coined the term 'integral psychology', discusses Sri Aurobindo's approach to the problem of the unconscious as comprehensive, evolutional and introspective. He maintains that in the Indian approach "yoga has been a necessary concomitant discipline for each system of philosophy for the realization of its truths and, therefore, the growth of personality is an indispensable issue for each system" (p. 2). He also adds that "'dwandwa' or duality and division are regarded as a general feature of mental life and that the aim of yoga is to create a unity out of it. And for this purpose 'Karma-Kshaya' or the dissolution of disposition is considered necessary" (p. 3).

According to Sen (n.d.):

to Sri Aurobindo the teleological or forward moving character is the central fact of our consciousness. It is the evolutional urge of life generally, which unfolds in the ascending scale of the animal species a progressive growth in consciousness. Therefore, the unconscious is the large evolutional base from which consciousness emerges. However, if the past is any indication, then it can be definitely affirmed that the goal of this long evolutionary march must be the attainment of a consciousness fully come to its own. That is to say when the unconscious has been reduced to the vanishing point and the human individual becomes fully aware of himself and capable of acting out of such awareness (p.6).

Yoga as the practical psychology of self-knowledge is the process of becoming increasingly conscious and the goal of integral yoga becomes the eventual conquest of the unconscious. According to Sen, "it involves a persistent looking within and a thorough self-exploration leading to complete self-discovery and self-integration" (p.7). The basic method of self observation used by Sri Aurobindo is called introspection which is an inner-directed empirical method of observing the contents of the psyche and personality. This " has to be raised to such a pitch of efficiency that self-observation becomes a constant and spontaneous inner activity and not a deliberate process of special occasions" (p.8). Introspection should be dispassionate and free of moral judgments and religious injunctions.

While many forms of psychological work attempt to help human beings become healthier by creating a harmonious balance between the unconscious and the conscious dimensions of personality, integral yoga and psychology aim at complete transformation of personality by making conscious the entire content of the unconscious mind. Sen points out that most forms of yoga strive to incorporate the higher unconscious into the conscious personality but only touch the surface of the unconscious for the purpose of purification of the topmost level of the unconscious from which contents surge up. Sri Aurobindo's approach, however, requires a complete investigation of both the higher and the lower unconscious realms—the 'subconscient'.

By the Subconscient Sri Aurobindo means the submerged part of the being in which there is no waking consciousness and coherent thought processes, will or feeling or organized reaction. Subconscient materials rise up into our waking consciousness as repetition of old thoughts and vital and mental habits and samskaras (impressions) formed by our past. There are three types of differentiation in the subconscient: the mental, the vital, and the physical subconscient, each one of which is distinguishable by the virtue of their contents and action on the waking personality. These subconscient processes are generally disorganized and chaotic. In other words, there is no execution of a unified will in the subconscient as the various impulses therein act chaotically and without any organization and thus various conflicts and struggles arise within the subconscient mind in addition to conflicts with the elements of our conscious personality related to the external environment.

Using methods such as hypnosis, free association, and dream analysis, Freud's therapeutic aim was to help the patient make conscious certain amount of the unconscious materials in order to create a balance between the conscious and unconscious mind. Sri Aurobindo was interested in much more than making the unconscious, passively conscious. Rather he was interested in the transformation of personality from the ordinary egoistic state to a fully conscious and integrated state.

Sri Aurobindo was careful, however, not to recommend plunging into the subconscient without first mobilizing the higher unconscious. Without this preparation there is a risk of losing oneself in the obscurity and the chaos of the Subconscient world. Sri Aurobindo's yoga

in the human being. Jung was also aware of the need for such a higher integrating principle which he termed the archetype of the Self, or the soul.

Depth psychologists first discovered the unconscious through their encounter with the pathological manifestations of the unconscious. Both Jung and Assagioli eventually realized the importance of the role of the soul (Self or Higher Self) as the catalyst for integration of personality, a task not possible through ordinary therapeutic techniques which often emphasize the importance of ego-strengthening which is necessary for those who suffer various forms and intensities of neuroses and psychotic dissociation, or

In integral yoga the goal is no less than the complete illumination, transformation and integration of the psyche and evolution of embodied consciousness.

is unique in that it starts with the opening of the higher centers of consciousness first. This is to avoid the trappings of the lower unconscious realms and intensification of attachments, as well as a myriad of other problems associated with premature opening of the kundalini energy without first establishing the Psychic will or even possibly Supramental will to guide the process of transformation of the unconscious. As Assagioli was also keenly aware, an important factor is the role of the will in this process. Unlike Freud's position, for Sri Aurobindo merely making the unconscious mind conscious is not sufficient for transformation and we need the assistance of the conscious will to help organize and transform the content of the Subconscient mind.

Another point of difference is that unlike Freud, for Sri Aurobindo the therapeutic aim is not to strengthen the ego. This is because ultimately the ego is self-centered even though it is better adjusted to reality and better organized than the impulses of the Id or the Superego. Therefore, a higher integrating center is needed which in integral yoga is the Psychic Being, or the evolving soul

even unmanageable phobias, depression or anxiety etc. Certainly for the initial healing phase strengthening the ego up to the point of health and stability is unavoidable and desirable. But when it comes to the complete transformation of personality as required in integral yoga and psychology, a mere balancing of the conscious and unconscious elements of personality through a healthy and strong ego will be insufficient.

Traditional depth psychology often focuses on expanding the sphere of human consciousness by incorporating materials from the unconscious regions to the conscious regions, while traditional yoga, attempts to engage with the higher realms of the unconscious or the transpersonal Self and is not necessarily interested in transforming the unconscious psyche as much as it is interested in developing the higher unconscious. This could result in disinterest in ordinary consciousness and evolution of embodied consciousness. In integral yoga the goal is no less than the complete illumination, transformation and integration of the psyche and evolution of embodied consciousness.

To summarize, the goal of yoga is to

accelerate the rate of conscious evolution. Integral yoga aims at total transformation of the unconscious as well as ordinary consciousness. Conscious evolution, therefore, requires a total transformation of the psyche and human personality. The high level of integration of personality required in this process supersedes the establishment of basic wholeness of personality which is possible by harmonizing the egocentric and psychocentric spheres of consciousness. This level of integration known as 'Psychic Transformation' in integral yoga and psychology, which is similar to Jung's process of 'Individuation' or integration of ego and Self, is a necessary foundation. The complete eradication of the unconscious however, would necessitate the activation of Supramental consciousness.

This is the evolutionary challenge of our time.

References

Assagioli, R. (1971). *Psychosynthesis*. New York: The Viking Press.

Assagioli, R. (1973). *The Act of Will*. New York: The Viking Press.

Chaudhuri, H. (1975). Yoga Psychology. In C. Tart, Transpersonal Psychologies (pp. 233-280). New York: Harper & Row Publishers

Chaudhuri, H. (1977). *The Evolution of Integral Consciousness*. Wheaton, Il.: Quest Books.

Jung, C. G. (1959). *The Collected Works of C.G. Jung: Aion* (2nd ed., Vol. 9, Part II) (R.F.C. Hull, Trans.). Princeton, NJ: Princeton University Press.

Meher Baba (1997). *God Speaks: The theme of creation and its purpose*. Walnut Creek, CA : Sufism Reoriented.

Sen, I. (n.d.). *The Unconscious in Sri Aurobindo: A study in Integral Psychology*. (Unpublished manuscript).

Sri Aurobindo (1970). *Letters on Yoga. Sri Aurobindo Birth Centenary Library* (Vol. 22). Pondicherry, India Sri Aurobindo Ashram Trust.

Sri Aurobindo (1992). *The Synthesis of Yoga*. Pondicherry, India: Sri Aurobindo Ashram Trust.

Sri Aurobindo (1997). *The Life Divine*. Pondicherry, India: Sri Aurobindo Ashram Trust.

Atemporal Creativity
Evolution Beyond Lines & Spirals

Zayin Neumann

Jean Gebser's work offers a vision of *time-freedom*, wherein the modern soul is able to free itself from the constraints of the world that is quickening all around us. Other contemporary authors offer Integral visions of their own, but it is in this time-freedom, this *anachronon* that Gebser shows his particular form of genius. This *atemporal* vision of reality is a direct critique of what he terms *biological and historical* theories of evolution. He states quite clearly that "the mutational process we are speaking of is spiritual and not biological or historical" (Gebser, 1985, p. 37). By this he means that a linear (historical) view of reality is a both a serious over simplification of the facts, and a very onerous mistake.

In its simplest form, evolution tends to be thought of as a linear developmental process. For Jean Gebser, "Chronological time is but one aspect of a more

Zayin Neumann, M.A. is a doctoral student at the California Institute of Integral Studies, where he is working at the intersection of various contemporary wisdom traditions, a process-oriented philosophy of science, and a deep intuition into the gifts offered by an evolutionary theory of consciousness. Zayin organizes and facilitates workshops on integral cosmology, holistic integration/sexuality, various forms of shamanism, and leads contemporary rites of passage. He has a private practice in the Bay Area. Contact: 2735 Elmwood Ave., Berkeley, CA 94705. 415.793.9683. Zayin@Freedomtoinhabit.com

encompassing phenomenon" (Gebser, 1985, p. 284). This chronological time is held in contrast to the various cyclical experiences of time that can be found throughout our indigenous wisdom traditions, where 'time' is closely aligned with the movements of nature, and has far less directionality inherent to it. There are also experiences like that of some of our nomadic ancestors (i.e. the aboriginal peoples of what is now called Australia) where time is a barely punctuated dream-time. Following the work of Gebser, we can easily deepen these basic insights into different 'times' by exploring the language and art of those people who fall within his Magic, Mythic and Mental epochs of human consciousness. Alongside these epochs Gebser also considers timelessness— a lived experience that has little or no reference to what we would consider 'time'. He also speaks to an *atemporal time-freedom* that is currently unfolding and rooting itself in actuality.

Rather than a linear model of evolution, Gebser speaks of time as irrupting, which he tells us is the "profound and unique event of our historical moment" (Gebser, 1985, p. 283). There is a leap or mutation which will not happen in (linear) time, but rather will concretize all the various lived experiences of time. Such a mutation will include

linear time, but will not in any way be limited to or by linearity. He tells us in an essay entitled 'Psyche and Matter' (Gebser, 1996) of time as a 'pre-existing whole.'[1] In *The Ever-Present Origin*, Gebser (1985) writes that the "coming to awareness of time" in its full complexity is a precondition for the awakening consciousness of time-freedom. The freedom from time, in turn, is the precondition for the realization of the integral consciousness structure that enables us to perceive the aperspectival world" (p. 289). Within the unfolding Integral mutation time itself has been realized and concretized, and the whole of time is free to express itself in the myriad of ways that is its nature. Three basic examples are of point-like 'Magic' time, cyclical 'Mythic' time, and linear 'Mental' time. Evolution, as an 'Integral' understanding, must accommodate and express the full richness of just such a time, without relying too heavily on linear Mental time.

To accentuate this time-freedom Geb-

[1] We must be careful here not to fall into Ken Wilber's pre/trans-fallacy. Gebser is not suggesting that Origin has always existed as a static whole. He is speaking to a phenomena akin to Whitehead's understanding of eternal objects and epochs of existence. wherein absolute freedom exists within a basically given set of possibilities. To put this another way, a child does not experience the same Origin as the adult mystic in that different possibilities are ingressed (made actual) within the experience of each.

ser points to Einstein's theory of relativity,[2] Picasso's cubism, and the words of Rilke, T.S. Eliott, and Jorge Guilleniv[3] as various expressions of this coming anachronon. There are countless examples that could be given of more contemporary expressions of this impulse. We will explore the work of these artists and thinkers here as this paper is less about what the coming mutation of consciousness will be experienced as, and more about the actual process of evolution itself. In order to account for such an atemporal experience of 'time' we must introduce a level of complexity into our thinking that is not available to us in a simple linear causal understanding of reality. We must ask the question, "How does Gebser's thought manage to honor an evolutionary intuition, while not falling into the linear developmental trap?" The answer can be found in the work of Alfred North Whitehead as he endeavors to illustrate how true novelty can come into the world.

Creativity
Whitehead's First Principles

True novelty, Whitehead tells us, is neither dependent on efficient causation, nor on linear development. This point is absolutely crucial to the intuition of an Integral epoch of which Gebser writes. If Whitehead can clarify exactly how novelty can arise in a non-linear, acausal way, then Gebser's work has far reaching and deeply pragmatic ramifications. The crux of the comparison between Whitehead and Gebser lies within Whitehead's consideration of propositions, but in order to get there we must first attempt a brief summary of three of his first principles.

To this end, Whitehead's metaphysical project is an attempt to come to a set

Freedom from time is the pre-condition for the realization of the integral consciousness structure that enables us to perceive the aperspectival world.

of general ideas that is sufficient to the task of interpreting the entirety of our lived experience. Whitehead is quite aware that this is no small undertaking. He is quick to inform us that it is in fact an impossible task for the human mind to enter into, and yet it must be done. Metaphysics, when seen in this light, becomes more of a practice, a way of living more akin to Gebser's verition (living-in-beauty) than to a new system capable of describing in axiomatic truths the way things are. Gebser tells us that the efficient form of the Mental epoch is an expression of directive, discursive thought. This as opposed to the deficient expression of the Mental described by Gebser as divisive, is immoderate hair-splitting. We can feel in the efficient form of the Mental a tendency towards the broadening of our horizons towards greater transparency. In this broadening towards transparency we see Whitehead's project, and essential to this project is a set of general ideas that are coherent, logical, and necessary for the task at hand.

These general ideas are logical in that they do not contradict one another. They are coherent in that they presuppose each other, and cannot be understood separate from one another. They are necessary in that there is some one thing that is, and that beyond which no relationships can be understood rationally. This one thing for Gebser is the 'Ever-Present Origin', and Whitehead informs us that

it is towards the essence of something similar to this Origin that his speculative philosophy strives. The three general ideas that Whitehead settles upon are *Creativity, eternal objects, and actual occasions*. I will attempt to define these three principles before moving on to a consideration of propositions.

Actual occasions tell us stories about what has been. For Whitehead they are the only 'really real things' in existence. They do not exist in isolation, but rather are the satisfaction of a subjective choice, predicated on the concrete experience of everything that has been actual, juxtaposed against everything that is both relevant and possible for the subject, or occasion, that is making a decision. This is a dense sentence, and it holds Whitehead's key to the very nature of actuality. The process of becoming 'actual' requires three phases of what Whitehead terms concrescence.

The first phase operates with an emphasis on the physical pole of life. Absolutely every single occasion that has actualized itself (actual occasion) tells the story of what has been by being felt as the concrete past actual world (physical prehension). The concrescing occasion literally feels the entirety of the actual universe. As it feels the entirety of the universe, it is also presented with an array of everything that exists in potential relevance to its past experience. Here it makes a choice (the second phase of concrescence) as to how to proceed. It can place its emphasis on the physical pole, and more or less bring the past into the present in a conformal fashion. It can also place more or less emphasis on the mental pole, thereby allowing a greater possibility of freedom and novelty in relation to its past. Once the occasion makes a decision it moves into the third and final phase of its process of becoming. It has a moment of actual existence in which it enjoys or appreciates the decision it has made. It then becomes the objective past of future actual occasions. In effect, the occasion is a process of feeling, considering/choice, and appreciation.

Inherent in this process of becoming, there is also the *subjective aim* of any

[2] It is interesting to note the long standing tension between the thought of Einstein and that of some of his contemporaries; most notably Neils Bohr. The inability of contemporary physicists to accommodate both relativity theory and quantum physics acts as another pointer in the direction of the mutation Gebser discusses. Einstein can be seen as a clearly modern thinker, while Bohr clearly of the postmodern camp. The struggle between them becomes a moot point within the unfolding Integral mutation, at least that is the assumption of this paper.
[3] In *Psyche and Matter* (1996) Gebser quotes many of these authors. Jorge Guillen is quoted as writing, "The events - where are they, when did they take place? There is no so called history. There was a glowing, and this still is glowing. A single day, deeply engraved in our recollection, entrusts itself to the eternal present." Gebser goes on to quote Arthur Eddington as saying, "Events do not happen; they are just there and we come across them. The 'formality of taking place' is merely the indication that the observer has on his voyage of exploration passed into the absolute future of the event in question."

concrescing occasion. Much consideration must be given to subjective aims within the context of relating the decisions and enjoyment of occasions to the movement of evolution. We must consider in depth the origination of particular subjective aims, and of larger ecological and societal (in the sense of Whitehead's *societies of actual occasions*) aims. I will touch on these issues when discussing propositions later in this paper, albeit only briefly.

Returning to the three phases of concrescence, there is absolutely no actual existence outside of such a process of becoming. Whitehead tells us "[t]here are not 'the concrescence' and 'the novel thing': when we analyze the novel thing we find nothing but the concrescence... abstraction from which there is mere nonentity" (Whitehead, 1978, p. 211). The world, then, is made up of a series of processes, rather than a diversity of discrete and isolated *things*. This is so important that it must be emphasized again. There is no static knower, no passive known, these terms are relative and actuality is one of process not substance. For Whitehead this is the great mistake of the Western philosophical tradition. "In the place of the Aristotelian notion of the procession of forms, [the philosophy of organism] has substituted the notion of the forms of process" (Whitehead, 1968, p. 140). This point can be very controversial. It is beyond the scope of this paper to delve more deeply into the subject, and yet we must assume this point if we are to continue in our endeavor to understand through Whitehead's lens how novelty can be introduced into actuality.

In order to better understand this introduction of novelty we must come to a working knowledge of what Whitehead terms eternal objects. Eternal objects, he tells us, are pure possibility. They are in no way connected to any definite set of actual occasions. They 'exist,' not in actuality, but in potentiality. The greater the emphasis that is placed on the mental pole and conceptual prehensions, the more possibilities or eternal objects

will become relevant. All eternal objects are possible, always, but for each actual occasion, only those eternal objects that are relevant to the particular occasion are recognizable. Relevance here is a matter of emphasis on physical and mental poles, in relation to past actualities, contemporary context, and subjective aim.

At one end, a complete emphasis of the physical pole would mean that absolutely no novelty would come into existence. The past would simply repeat itself *ad infinitum*. On the other extreme, a complete emphasis on the mental, absolutely all novelty would be possible. Nothing would ever "actually" exist, as

The world is made up of a series of processes, rather than a diversity of discrete and isolated things. There is no static knower, no passive known, these terms are relative and actuality is one of process not substance.

everything would always be possible, in effect disallowing the process of decision. Both of these extremes are hypothetical, because in our lived experience we experience ourselves as striking a balance of emphasis between the mental and the physical. Complex occasions emphasize the mental for particular durations, but must test these adventures against the test of ingression. They must bring their emphasis back to the physical pole, with a decision to actualize the felt possibility. As more complex occasions begin to influence less complex (occasions that have heretofore placed limited emphasis on mental pole) occasions, a greater variety of eternal objects can be ingressed in actuality. If less complex occasions influence more complex occasions (a kind of habituation by peer pressure) with regard to what is possible, less variety is possible. This dance appears to be essential to the very nature of reality. In order to understand actual occasions and eternal objects we

must attempt to penetrate the mystery of this dance between physical and mental polarities.

We can now approach Whitehead's third principle, 'Creativity'. We have come to understand that actuality requires a decision, whereby what has been chosen can be experienced. Without a decision, there is no actual occasion. The greater the freedom of choice that is possible, the greater the variety of eternal objects that can be ingressed or experienced. The question arises, "Why experience or actualize an ever larger number of eternal objects? Why is there novelty at all?" When Whitehead looks at his own experience, he sees a tendency towards greater complexity, freedom, and novelty. This is his experience, and so his metaphysics must account for this. He sees Creativity, *the many become one and are increased by one*, as the basic impulse of the Kosmos. Inherent in the very structure of reality is an expansion towards novelty. By including this third foundation, he can account for not only the *how of becoming* (actual occasions concrescing), and the what of becoming (eternal objects ingressing), but the *why of becoming* (Creativity) as well.[4]

---
[4] The reader might notice that at this point the when of becoming has not been discussed. This is an important issue, and brings up a fourth foundational principle. As Creativity is a movement towards conjunction and novelty, there must be a counter movement towards disjunction and entropy. An overemphasis on Creativity could be similar to an overemphasis on the stage of evolution, at the disregard of the stage of devolution (the natural death and decay inherent to actuality). It is beyond my own understanding of Whitehead to discuss in any detail how he would address this issue, but the issue itself is instructive concerning our topic. If there were only evolution, only an emphasis on the mental pole, the when would be a moot issue. The entirety of actuality would realize itself as everything that is possible, and nothing at all interesting would happen. There would simply be all that is possible. The when requires the dance of the physical and mental poles. For actual occasions to exist as unique individualities, not having realized themselves as everything that is possible, there must also be some principle at play that restricts their ability to realize all eternal objects at once. I will call this principle Soil. I cannot go into any great detail regarding Soil here. This is because I do not know exactly how Whitehead deals with this issue.

We have only just touched upon what is possible in considering Whitehead's metaphysics. In our own small way, we have added to his incredibly successful project meant to render the essence of *becoming* transparent. We could spend several years and many papers delving more deeply into these three principles, but for now we will move on to Whitehead's consideration of propositions.

Propositions

Our purpose within this essay is to account for the introduction of non-local acausal novelty via the activity of evolution. In order to effectively deal with Whitehead's answer to this riddle we have introduced the three first principles mentioned above, namely 'actual occasions', 'eternal objects', and 'Creativity'. Having done this we can now move to the focus of our consideration of what Gebser terms spiritual evolution by looking to Whitehead's rendering of propositional feeling, propositional thinking, and rational knowing.

Evolution is clearly one of the greatest discoveries of the Integral mutation of consciousness. It is through the dawning realization of an evolutionary stage of becoming that we begin a heretofore unheard of self-reflexive act. It is in the dawning of evolutionary thinking that we begin to look back on ourselves and penetrate deeply into the *how*, the *what*, and the *why* of becoming. We also begin to know the *when* and the *where* of becoming. In the Integral mutation both time and space are set free, no longer subject to the measurement of stick and clock. I would like to suggest that by so completely teasing out the how of evolution, Whitehead has managed to accomplish in his own way the new freedoms of 'diaphaneity' and 'verition' of which Gebser speaks. His explication of propositions is the fullest understanding of the means by which freedom and novelty are introduced into any system that I have come across. I will proceed now to delve into his careful consideration of this process.

Whitehead defines propositions as stories that lure us (actual occasions) into novel futures. When I create the sentence, "I intend to not only to understand, but directly participate in evolution so as to grow a living house," I am positing a novel future. This sentence is a proposition, and it is through such propositions that Whitehead's foundational impulse of the Kosmos, Creativity, can ingress greater freedom into actuality. To put this more simply, proposi-

tions tell us what might be, "the sun will rise tomorrow," and lead us towards such futures.

By placing emphasis on the conceptual prehension phase (mental pole) of a concrescence, the occasion that is becoming moves away from Soil,[5] towards Creativity, in order to become aware of eternal objects that have not yet been ingressed in actuality. In simple occasions this phase of concrescence is satisfied by bringing the past into the present through a decision to maintain the past conformally.

Something very important happens in more complex occasions, whereby greater and greater emphasis is placed on this conceptual phase. The proposition, the statement acting as lure, enters into the awareness of the concrescing occasion, effectively becoming a new datum or object to be prehended. The proposition itself contains three basic parts. "I will grow a living house." In this proposition, there is a definite set of past actualities that is being physically prehended. There is a particular "I" that is being referenced (This proposition is far less relevant to a door, or a dog, or a grandmother in Omaha baking cookies, than it does to the author of this paper (growing houses is relevant to the particular society of occasions that is "me"). The "I" is made sensible by being relevant to a definite set of past actualities. This 'I' becomes the *logical subject* of the proposition.

Also included within this sentence are the words living and house. These two words point to a definite set of possibilities that are conceptually prehended, relative to the logical subject of the sentence. A living house could mean a tree, if the "I" in the sentence is a bird in a child's story that is talking about its house. In the sentence above, assuming the "I" is the author of this paper, the words "living house" take on a very different meaning than the one found in the children's story. These words make

[5] See note 4.

Art: Salma Arastu

up a *predicate*, meaning that they consist of a definite set of possibilities that are relevant to the logical subject of the proposition.

There is one more important part of the proposition above, which is the propositional feeling (also called a copulative verb). The logical subject means to integrate or actualize the predicate of the sentence. In this case, "I" mean to "grow a living house". The propositional feeling points to the degree to which the predicate corresponds to the logical subject. "I have not grown a living house," has a greater correspondence between predicate and subject than, "I am growing a living house." We can see here that propositions can tend towards truth (conformal propositions) or falsity (non-conformal) propositions.[6] We are not too concerned here with conformal propositions. Such propositions simply tell us what has been the case in the past; they do not lure us towards novel futures. "[F]alse propositions have fared badly, thrown into the dust heap, neglected. But in the real world it is more important that a proposition be interesting than that it be true" (Whitehead, 1978, p. 259). Non-conformal propositions are our vehicles of the "creative advance," the mechanism of evolution. It is non-

discussing for the remainder of this section.

Evolution requires a modicum of freedom wherein there are relatively few restrictions on what is possible. This means that a proposition like 'the car is powered by gas' provides less opportunity for evolution than the proposition 'the car is powered by water.' Evolution requires freedom from the past,

Evolution is clearly one of the greatest discoveries of the Integral mutation of consciousness.

and freedom from the past requires an increasing intensification of emphasis on the mental pole. Whitehead breaks propositions into several stages of increasing mental intensity. He is clear that his designation of stages is somewhat arbitrary in that they could be distinguished in an infinite number of ways. Following the lead of Thomas Hosinki (1993) we shall work with the three basic stages of *propositional feeling, intellectual feeling, and rational knowing*.[7]

Whitehead tells us that the first of these three, propositional feelings, indi-

decision making processes with the way Whitehead is using the word here, but that would be a mistake. There are many ways in which we make 'decisions.' If we largely ignore all future possibilities, then we generally bring the past into the present in a conformal manner. When we do this, we are primarily working with the 'true' or conformal propositions we mentioned above. Propositional feelings are one step towards the mental pole from these conformal propositions, because the proposition that is being felt is not true but false. It is pointing to a possible future, not an actual one.

We say that a propositional feeling is still relatively unconscious because even though the contrast between actual and possible is felt, the result is an unconscious reaction. If I am walking along saying to myself, "man, man, man" and this is true, then we are working with a conformal proposition, and no novelty is introduced. If I am walking along saying to myself, "human, human, not human, human", then some novelty has been introduced. Maybe as I am walking along I suddenly turn left towards my home, moving unconsciously towards a place where I know I can find some reassurance that I am indeed a human. The way this example is written is important. I did not write, "I am human," but rather, "human." I wish to point out with this sentence a contrast that is felt between human and not human, while a certain level of non-reflectiveness is maintained. We have introduced a false proposition, not human, within a context where the conformal proposition is actually "human." A certain amount of emphasis is placed on the mental whereby a false proposition may be felt, while the action that is taken (turn left) is largely unconscious. This is an important point, which we will return to in the next section. There is just enough mental intensity in a propositional feeling whereby a possible novel future can be entertained, but not enough so that a

Gebser suggests to us that evolution is atemporal and acausal, that it occurs via mutations or leaps that do not fall within linear or causal models of understanding.

conformal propositions that we will be

cates the feeling of a contrast between what is possible and what is actual. Once this contrast is felt, the subject (concrescing occasion) reacts to this felt sense in itself, and proceeds to make a relatively unconscious decision. If we remember back to our breakdown of an actual occasion, we notice every concrescence has three phases. It feels the concrete past, looks into the possible future, and then makes a decision based on these two prehensions. We can easily make the mistake of conflating our own

[6] It is important to note a shift from traditional Aristotelian logic to a kind of propositional logic. Where Aristotelian logic focuses on what is similar by way of asserting categories of being, we could say that propositional logic looks for what is different or interesting. In this way we begin to focus on process rather than content. To say this another we way, we find ourselves looking for that unifying process whereby what we are considering becomes interesting and useful within our thought. This unifying process is not a sum of parts, but rather a play of "parts." To use Whitehead's words, we are looking for higher grade occasions that are influencing the lower grade occasions which we are considering. For a lucid account of this shift of logics see chapter one of Ernst Cassirer's *Substance and Function* (Chicago: Open Court Publishing, 1923).

[7] Rational knowing is a term coined by Hosinski (1993) that is helpful for our purposes. It is not used by Whitehead in this particular way.

conscious action can actually be taken to ingress this possible future. Only an unconscious action is possible at this point.[8]

As we increase the emphasis of mental intensity we move to the next phase of propositions, which Whitehead calls intellectual feeling. During an intellectual feeling an awareness of the contrast begin to understand what Whitehead is attempting to explain to us. He is saying that consciousness enters into actuality at the point where a contrast is felt within the duration of one event. This awareness is an awareness of relationship between an actuality and a possibility, and it requires of us to slow down, to place emphasis on the mental pole a difference is felt, or becomes relevant within a moment. There is no "I" that recognizes this difference over the course of several moments, but rather consciousness is constellated around difference for a moment and then this moment passes into another. Within an intellectual feeling a contrast is felt. We call is a contrast rather than difference, because within one moment two polar opposites/complementarities are felt, but not in opposition or separation from one another. What is felt is not one side or the other of the contrast, but rather the contrast itself. Notice also that no "I" maintains awareness of this contrast from one moment to another. The contrast occurs within a moment of experience, that passes away into another experience. It is only in the final stage of concrescence that an I or an awareness is maintained across moments. Our contrast in a moment, becomes a kind of either-or over the course of several moments. There is an intensification along the mental pole to the extent that awareness is maintained at least as long as it takes to compare one proposition with another. Rather than actual datum, non-conformal propositions become the focus of attention. Rather than focusing on what is actual, we have moved to a point where what is merely possible becomes relevant and the focus of our awareness.

It is in the dawning of evolutionary thinking that we begin to look back on ourselves and penetrate deeply into the how, the what, and the why of becoming. We also begin to know the when and the where of becoming.

between what is possible and what is actual begins to take shape. This allows the formation of a judgment, which opens the possibility of a conscious decision. If I am walking along saying to myself, "human, human, human-not human, human, human, human-not human," something new has entered into possibility. Notice the difference between, "human, human, not human, human," and, "human, human, human-not human, human." In the first, something different comes into the awareness of one moment, whereas in the second a contrast is felt between what is and what might be within one experience. In order for a contrast to come to "awareness" or consciousness it must happen within the same event. We can see this even more clearly when we look at the words "not human" by themselves, and then look at the words "human-not human" by themselves. The second actually requires of us to pause for a moment and consider the relationship that is present. This is not the case with the first set of words.

By feeling into this difference, we can

for a greater duration before coming to a decision. We are being asked to make a conscious judgment. This may seem like a simple point, but it is absolutely crucial for the discussion at hand and should not be overlooked. Whitehead is pointing out to us the very point at which consciousness enters into a concrescing occasion. In effect he is pointing at the mechanism whereby consciousness (and novelty) enters into actuality.

If we take even more time emphasizing an increased intensity of the mental pole, we will come to experience that we are here calling rational knowing. Rational knowing comes into play when an occasion is able to compare and contrast over the course of several moments. In this way, a concrescing occasion can assess the value or the truth of the contrasts that are felt. One example of this would be to take a non-conformal proposition, "I am not human," and compare it with a conformal proposition, "I am human." This can be done over the course of several moments, so while actually feeling human, one can compare the possibility of not being human to the conformal feeling of being human.

Notice what an incredible difference there is between the use of these propositions and how propositions were used within propositional and intellectual feelings. In propositional feeling

Within this phase of rational knowing an occasion can be so self-reflective as to consider an obviously false proposition over and against the actuality of its experience. We have now opened up the possibility of truth statements. An occasion can differentiate between human and not human, and decide whether or not what it is feeling itself as is human or is not. In effect, we have defined human. We can now compare "human", with other things that are "not human". Not-human might be one example, but so is tree, or skateboard. We can compare experiences that have little or no obvious relation to one another, because an emphasis has been placed on the mental pole to the extent that several propositions can be compared within the experience of one event or occasion. In addition to such truth statements, and similar value statements, notice that the occasion has recognized itself as exist-

---

[8] I like to use the words Iother written in this way to show the lack of differentiation. As the reader continues they might find it interesting to consider intellectual feelings as an I-Other, where an emphasis is placed on the hyphen. When moving to rational knowing, we can start to see the relationship as , I Other, where the hyphen has gone, and an "I" stands over an apart from the "Other."

ing over the course of several moments, and so is also able to coherently say I am human. Up until this point an "I" statement is not within the realm of actuality, but it can now be considered. We have moved from feeling a contrast and reacting in an unconscious way, to becoming aware of a contrast and making a conscious judgment, to becoming aware of a subject (an "I") that exists over several moments (and subsequently an objective world out there).[9] This has all happened because of an increased emphasis on the mental pole of an occasion.Here an occasion can be so self-reflective as to consider an obviously false proposition over and against the actuality of its experience. We have now opened up the possibility of truth statements. An occasion can differentiate between human and not human, and decide whether or not what it is feeling itself as is human or is not. In effect, we have defined human. We can now compare human, with other things that are not human. Not-human might be one example, but so is a tree, or a skateboard. We can compare experiences that have little or no obvious relation to one another, because an emphasis has been placed on the mental pole to the extent that several propositions can be compared within the experience of one event or occasion. In addition to such truth statements, and similar value statements, notice that the occasion has recognized itself as existing over the course of several moments, and so is also able to coherently say I am human. Up until this point an "I" statement is not within the realm of actuality, but it can now be considered. We have moved from feeling a contrast and reacting in an unconscious way, to becoming aware of a contrast and making a conscious judgment, to becoming aware of a subject (an "I") that exists over several moments (and subsequently an objective world out there). This has all happened because of an increased emphasis on the mental pole of an occasion.

As we have moved from propositional feeling, to intellectual feeling, and on to rational knowing, we have increased the intensity of emphasis placed on the mental pole. By placing this emphasis on the mental pole, we have opened up greater and greater possibility for novel propositions and futures to be realized. If an occasion were to maintain this emphasis on the mental, then the more extraordinary propositions that may have been glimpsed would remain as potentialities only. The furthest reaches of the mental imagination will remain mental imaginations as long as emphasis is kept at this far end of the mental pole. In order for extraordinary propositions to embody themselves in actuality emphasis must shift back to the physical pole of life.[10] We can see this in a person who tends towards flights of imagination. If this person is never compelled to test their imaginations against the physical limitations of actuality, then they simply wander further and further off into imagination. As emphasis on the mental continues in this way, the imaginations, or non-conformal propositions, begin to lose any reference to the physical. In effect, the predicate of the proposition has less and less relevance to the logical subject of the proposition, and so the copulative verb becomes meaningless, as the concrescing occasion cannot integrate the ever more disjunctive pieces of the proposition. Without an attempt to check the possibility of ingressing the proposition into the actual, the occasion gets lost in a world of phantasy.

Having reached the limits of non-conformal propositions that have the possibility of becoming actual, we must shift the emphasis back to the physical pole

so as to attempt to involve or actualize these non-conformal propositions. Here

Many contemporary theorists within the fields of Integral studies have aligned themselves with a more or less linear understanding of time, while various traditional and indigenous authors and thinkers tend towards more cyclical experiences of time.

we see an end to the evolutionary stage of becoming, and a return to the involutionary stage. The seeds of possibility that have been realized must now be tested against the past actual occasions.

Back to Creativity

We started this section by stating that true novelty is neither dependent on efficient causation nor on linear development. Our purpose on one level has been to call into question those theories and systems that put forward an evolution of simply located "things" through a progressive linear development. Our intention is to illustrate a view of evolution that honors the incredible complexity of what Gebser has termed spiritual evolution. By way of attempting to meet the complexity of an evolution by leaps and mutations we have delved into the collective writings of Alfred North Whitehead, paying especially close attention to the 4th and 5th chapters (pp. 256-280) of Part III of *Process and Reality* entitled the *Theory of Prehensions*.[11] Our goal has been to clarify exactly how novel mutations of consciousness could enter into actuality, and in the process we have hopefully shown how consciousness itself enters into actuality.

Whitehead is deeply committed to the notion of ultimate freedom. From this starting point he begins to unfold a view of reality that is non-substantive in nature. To this end he is highly

---

[9] We might say that the mind-body problem of "out-there" is a deficient Mental problem within Gebser's illumination of human mutations of consciousness. It can also be seen as a seed for the dawning Integral epoch. This is a subtle point beyond the scope of this short paper.

[10] It is my own assertion that the Integral epoch will require a shift from emphasis on the mental pole to an emphasis on the physical pole. So much novelty has been introduced (evolution) in our lives, that in the coming years there will be an emphasis on what I would call involution (not to be confused with Sri Aurobindo's use of the word), rather than evolution. I find the work of Dane Rudhyar (1970) to be of particular interest here.

critical of what he sees as an Aristotelian tradition that emphasizes a procession of forms. By inquiring into the forms of process, Whitehead's work speaks to how something truly novel can come into existence against the backdrop of Creativity. Viewed in abstraction an object appears to be passive, but when we take a closer look we see that there is no such static thing. What we find, says Whitehead (Whitehead, 1967, p. 179), is an active process of creation whereby what is novel comes into the world. "Creativity is the universal of universals characterizing ultimate matter of fact. Thus creativity introduces novelty into the content of the many... the many become one and are increased by one... Thus the 'production of novel togetherness' is the ultimate notion embodied in the term 'concrescence.'" (Whitehead, 1978, p. 21). In considering Whitehead's words there can be no mistaking the importance of Creativity within his thought.

Based on this non-local reality, Whitehead allows himself an incredible freedom to consider multiple forms of causality. In a traditional Newtonian world, we are limited by an efficient causation that requires a simply located object to effectively run into another such object by way of producing a causal relationship.[12] Such simple passive objects are for Whitehead dead and lifeless. "Thus for Newtonians", he tells us, "nature yielded no reasons: it could yield no reasons. Combining Newton and Hume we obtain a barren concept... it is this situation that modern philosophy from Kant onwards has in its various ways sought to render intelligible. My own

belief is that this situation is a reductio ad absurdum, and should not be excepted as the basis for philosophic speculation" (Whitehead, 1968, p. 135). These are strong words, and he goes on to ask, "How do we add content to the notion of bare activity? Activity for what, producing what, activity involving what?" (Whitehead, 1968, p.147). His answer? "Process for its intelligibility involves the notion of a creative activity belonging to the very essence of each occasion. It is the process (of concrescence of actual occasions) of eliciting into actual being factors in the universe (eternal objects) which antecedently to that process exist only in the mode of unrealized potentialities" (Whitehead, 1968, p.151).[13] We could say that this is the how and the why (Creativity) that novelty (unrealized potentialities) are ingressed in actuality. Creativity, then, is the short answer to the question that we first endeavored to ask with regard to Gebser's intuition of spiritual evolution over and against what he termed biological and historical evolution (and what we might call an evolution of lines and spirals).

Atemporal and Creativity

Evolution, I would suggest, is an Integral seed that has taken root by way of catalyzing the momentous leap of consciousness authors like Jean Gebser, Teilhard de Chardin, and Sri Aurobindo assure us is more than a passing pos-

sibility. Following Gebser's lead, it is increasing crucial that any contemporary dialogue with regard to evolution take very seriously questions about the nature of time. To this end, many contemporary theorists within the fields of Integral studies have aligned themselves with a more or less linear understanding of time, while various traditional and indigenous authors and thinkers tend towards more cyclical experiences of time. For those in the first camp, evolution becomes a developmental, progress-oriented march into the future, while for those in the latter camp, evolution is something of a misnomer, as to them it is obvious that we as a species are moving away from a golden era. For this second group it is clear that we need to consider the fact that we have lost our way and need to return to that golden age. There are still others who one encounters in popular culture who raise the flag of a kind of timelessness, or an always now, or some such variation on the theme.

Evolution needs to account for the various wisdoms that are represented here, and the numerous others that await our respect and attention. This is often not the case, as evolution is generally aligned with an atomistic, linear, and causal experience of reality. This leads to developmental and historical models of evolution, as well as to theories of evolution by way of biological specialization. The primary purpose of this paper has been to extract evolution from such narrow and restrictive modes of thought. Gebser suggests to us that evolution is atemporal and acausal, that it occurs via mutations or leaps that do not fall within linear or causal models of understanding. Where Gebser points us in the right direction, I have found that it is crucial to look outside his own writing for a more thorough explication of this process. For this, I have turned to Alfred North Whitehead.

Whitehead (1978), in *Process and Reality*, makes explicit the function or the process of becoming whereby novelty comes into existence. In effect, Whitehead clarifies the actual mechanism of evolution. This is something that Gebser only hinted at, and that many other integral authors have wrestled with.[14] In order to accomplish this task, Whitehead puts forward a philosophy

---

[11] Thomas E. Hosinki (1993) suggests that very few guides are available for these complex chapters offered by Whitehead (p. p. 127n11). I have found Hosinki's work, Stubborn Fact and Creative Advance: An Introduction to the Metaphysics of Alfred North Whitehead (New York: Rowman and Littlefield Publishers, 1993), pp. 99 - 127, to be a place to start. Hosinski points towards Donald W. Sherbourne, A Whiteheadian Aesthetic (New Haven: Yale University Press, 1961), pp. 55 - 69; and Elizabeth M. Kraus, The Metaphysics of experience: A companion to Whitehead's "Process and Reality" (New York: Fordham University press, 1979), pp. 117 - 124. Both of these works can be useful. David Ray Griffin is also helpful on this point, Reenchantment Without Supernaturalism: A Process Philosophy of Religion (Ithaca: Cornell University Press, 2001), pp. 320 - 350. In the end, however, we must return to Whitehead's own work in order to find clarity within these most rewarding few pages (1978, pp. 256 - 280).

[12] The history of contemporary science quickly points to the limitations of such a materialist view of reality. There are countless new sciences, from physics to biology to sociology that call into question such a simple system. For his own part, Whitehead was deeply influenced by the new sciences of relativity theory and quantum theory of the early 1900s. If we start to follow the evolution of the new sciences we start to see a picture rife with struggle and conflict among various authors and thinkers. Einstein and Bohr went from being great friends to finding themselves barely able to stand in the same room as one another as their careers progressed over the years (Peat, 2002). I mention these struggles by way of emphasizing the very radical nature of Whitehead's contribution to these conversations. I would suggest that Einstein and Bohr struggled so because they were both holding on to the traditional Newtonian system that arose out of the Western version of Gebser's Mental epoch of human consciousness. Einstein stayed the course, insisting upon simply located things. Bohr attempted to deconstruct these 'things' through his non-local causation, but failed to step into the radically different world of actual occasions offered by Whitehead.

[13] Additions within parenthesis are mine.

of organism, that is non-substantive in nature (process oriented), that is based on what he suggests is the most fundamental property of reality. Creativity. He defines Creativity as "the many becoming one, and being increased by one" (Whitehead, 1978, p. 21). Gebser tells us that the evolution he speaks of is an evolution of *plus mutations* (Gebser, 1985, p. 38). Such an evolution is one of over-determination, whereby an openness to possibility is maintained. As we have seen this notion of "openness to possibility" is akin to something Whitehead proposes through his discussion around propositions and the higher phases of concrescence.

Whitehead's metaphysic, then, gives us a lucid and coherent account of how novelty, leaps, and mutations can align with an atemporal acausal intuition into the nature of reality. Both Whitehead and Gebser are careful to make it as clear as possible that they are not speaking of linear developmental models.

A Footnote

By way of ending I would like to open the door for future discussions around comparisons of these two authors. Hav-

___

[14] The evolutionary philosophy of Sri Aurobindo offers an entirely different and unique description with regard to the process of evolution than that put forward here by Whitehead. Ken Wilber is another example of an important and unique voice within the field, one that tends towards a Mental lens, but which can offer a variety of unique insights.

ing made explicit Gebser's need for a clarification of the how's and why's of spiritual evolution, we have spent the majority of this paper considering Whitehead's account of the creative impulse that runs through the Kosmos. Having accomplished this task to a greater or lesser degree, future discussions might revert focus back to Gebser and his intuition into the story of how human consciousness has come to its contemporary (and potentially Integral) expression. I see an interesting and striking conversation unfolding at the intersection of Whitehead's understanding of what he terms the "higher phases of concrescence" and Gebser's description of the various epochs of human consciousness.

Upon close examination Gebser's mutations of human consciousness seem to express themselves as approximate corollaries of Whitehead's understanding of *concrescence* (the process of becoming) in the form of *conformal concrescence* being related to the Archaic epoch, propositional feeling related to the Magic epoch, intellectual feeling related to the Mythic epoch, and rational knowing seeming to express clear similarities with Gebser's Mental epoch. If we are able to outline the similarities inherent to these two thinkers, then we might just find ourselves with a logical, coherent, and necessary description of exactly how consciousness (that most novel of novel "things") came into actuality. Such a

comparison will surely offer itself as a fascinating point of departure for future conversations related to the work of these two authors, but one that will have to be saved for another day.

References

Gebser, J. (1985). *The Ever-Present Origin*. Athens, Ohio: Ohio University Press.

Gebser, J. (1996). Two essays: "The conscious and the unconscious" & "Psyche and matter.". *Journal of Culture and Consciousness*. 3, 86-91

Hosinki, T. E. (1993). *Introduction to the metaphysics of Alfred North Whitehead: Stubborn fact and creative advance*. Lanham, MD: Rowman & Littlefield.

Whitehead, A. N. (1933). *Adventures of Ideas*. New York: Free Press.

Whitehead, A. N. (1968). *Modes of Thought*. New York: The Free Press.

Whitehead, A.N. (1978). *Process and Reality*. New York: Free Press.

Cassirer, E. (1923). *Substance and function*. Chicago: Open Court Publishing.

David Ray Griffin, D. R. (2001). *Reenchantment without supernaturalism: A process philosophy of religion*. Ithaca: Cornell University Press

Kraus, E. M. (1979). *The metaphysics of experience: A companion to Whitehead's "Process and Reality"*. New York: Fordham University Press

Peat, F. D. (2002). *From certainty to uncertainty : The story of science and ideas in the twentieth century*. Washington, D.C: Joseph Henry Press.

Rudhyar, D. (1970). *The plantetization of consciousness: From the individual to the whole*. Harper & Row: NY.

Sherbourne, D. W. (1961). *A Whiteheadian aesthetic*. New Haven: Yale University Press.

The Oneness (and One-ing) of the Way

Using Both Hemispheres of the Global Mind

Sheri Ritchlin

In the late nineteenth century, voices were being raised simultaneously in China and in the West that were heralding a change larger than nations or even continents. On both sides of the planet, worldviews were opening up as if to anticipate and accommodate a world that would be both creatively and disastrously flung wide open through global wars, space exploration, the dramatic advances of modern communication and transportation, and through the complexities of a global economy.

As Alfred North Whitehead and the burgeoning generations of holistically oriented systems thinkers were leaning in the direction of an organic model of the universe resonant with the ancient Chinese cosmos, so too were Chinese thinkers seeking to arouse their country out of its long isolation and stagnation, leaning in the direction of the remodeled, post-enlightenment institutions of the West (Cheng, 2002). We can now see the two cultures as mirror images, developing from the unique tal-

ents of their separate hemispheres, moving slowly toward one another, while rediscovering the seeds of their own native genius. Each has entered the agonizing struggle to integrate the two hemispheres in a higher resolution and a richer global consciousness.

From the earliest periods of Chinese thought, a capacity for the union of Heaven and Earth was not only the uniquely human gift, but also the essential role of human consciousness in the cosmic process. The key to understanding the role of an integrative consciousness in Chinese thought lies in the philosophical meaning of the number "one," *yi* (also transliterated as *i*, depicted as a single straight line), which is Tao[1] within function as the action of the phenomenal world (Wu, 1990). As in English, the word *yi* is a noun or an adjective, but in this sense, as it unfolds into time, it is "one-ing."[2] Tao is the great Way of the universe, described by Lao Tzu as "beyond name and form," the mysterious mother of all

things, which comes into manifestation through this singular movement. "The way (Tao) of Heaven and Earth can be declared in one sentence," says the Confucist *Doctrine of the Mean*,[3] "they are without doubleness, and so complete in a manner that is unfathomable" (Legge, 1971, p.420).

The earliest and most complete description of the process that I have called one-ing appears in the Great Treatise on the *I Ching*—the *Ta Chuan*—that forms the Fifth and Six Wings or commentaries on that work, attributed to Confucius (550-479 BCE) but probably appended later, in the third century BCE. The *I Ching*, or "Book of Changes," was a system of sixty-four six-line figures used for oracular divination, traditionally dating back to the legendary founder of Chinese culture, Fu Hsi, around 2800 BCE. It is given its first written texts around 1000 BCE by King Wen and the Duke of Chou, the founders of the Chou Dynasty. The fourth of the *I Ching* sages in this tradition is Confucius, credited with the authorship of the Ten Wings or commentaries, which develop its philosophy.[4]

Sheri Ritchlin, Ph.D., is a free-lance writer and lecturer. She is the author of *One-ing* and *Dream to Waken*, as well as articles published in *Parabola Magazine, Noetic Sciences Review (Shift), The Evolutionary Epic: Science's Story and Humanity's Response*, and *The Ecozoic: A Tribute to Thomas Berry*. She is currently at work on *Fields of Light: 2012 and the Venus Transit of the Sun*. Her website is www.SheriRitchlin.com. Email: sritchlin@cs.com

[1] There are two systems of transliteration of Chinese characters, Wade-Giles and Pinyin. While the latter system is more currently in use, I have preserved the older forms of Chinese words in transliteration because they are more recognizable to the average reader and appear in the widely used Wilhelm/Baynes version of the *I Ching*.
[2] It should be noted that Tao itself has no direct action. Lao Tzu expresses this in Chapter 48 of the *Tao Te Ching* with the paradoxical statement that the Tao does nothing yet nothing is left undone.

[3] I use the word "Confucist" to identify those works traditionally ascribed to Confucius, but now regarded as being of a later date, although still hallmarks of Confucian thought. *The Doctrine of the Mean* is considered one of the classics of Confucius and was probably composed by his grandson.

The complete text, including the Ten Wings, has been described by JeeLoo Liu (2006) as the single most important work in the history of Chinese philosophy and the very foundation of Chinese culture. "Every Chinese person, with or without philosophical training, would naturally be inclined to view the world the way the [*I Ching*] depicts it" (p.26). The work is rich with insights into the integrative process of Heaven-and-Earth (the universe) as it awakens within the human. The Eighth Wing, or "Discussion of the Trigrams," opens with these lines:

> In ancient times the holy sages made the Book of Changes thus: They invented the yarrow-stalk oracle in order to lend aid in a mysterious way to the light of the gods. They put themselves in accord with tao and its power [*te*: also translated as "virtue"], and in conformity with this laid down the order of what is right. By thinking through the order of the outer world to the end, and by exploring the law of their nature to the deepest core, they arrived at an understanding of fate [destiny].[5] (Wilhelm, 1977, p. 262)

The thinkers who devised the system of sixty-four hexagrams were keen observers of nature and no less rigorous in their self-exploration "to the deepest core." China's first sage-kings, as presented in the Book of History, were exemplary models of this. The first of these is Yao (given dates of 2357-2256 BCE), who is said to have sent his brothers to the four corners of his kingdom, "in reverent accordance with their observation of the wide heavens, to calculate and delineate the movements and appearances of the sun, the moon, the stars, and the zodiacal spaces, and so to deliver respectfully the seasons to the people" (Legge, 1998a, The Canon of Yao section, Para. 3). Through observing the movements of light and shadow on the earth, the movement of stars in the heavens, and the changing habits of birds, beasts and people, Yao "set the calendar in order and made the seasons clear," (Wilhelm, 1977, p.190) exactly as we find it in Hexagram 49 of the *I Ching*.

Yao is described as reverent, intelligent and thoughtful, "regulating and polishing the people of his domain." Through his love, we are told, the nine classes of his kindred became harmonious and "the myriad states of his empire were brought into universal concord" (Legge, 1998a, The Canon of Yao section, Para. 1).

One-ing as the production of individual things

The description of the first sage-kings in the Book of History (ca. 6th century BCE), whether fact or legend, would make an indelible imprint on the Chinese people through the coming millennia and give concrete meaning to such language as "They put themselves in accord with Tao and its power [virtue], and in conformity with this laid down the order of what is right." (Wilhelm, 1977, p. 262). The Tao of Heaven-and-Earth was a thing to be trusted, without duplicity, and thus a source as well of the moral structure of the people.

Unlike modern cosmologists, the ancient sages of the *I Ching* were not attempting to explain the origins of the physical universe. They did, however, presume a unity out of which all things arise. Our own creation story, both scientific and Judeo-Christian, has a single temporal origin as the Big Bang or the creative act of God "in the beginning." What we find in chapter I.11 of the *Ta Chuan* is a description of an archetypal process by which all things come into being.

> Therefore in change there is the great ultimate [*t'ai chi*]. This is what generates the two modes [the yin and yang]. The two basic modes generate the four basic images, and the four basic images, generate the eight trigrams. (Lynn, 1994, p.65)

This is the first text in which the phrase *t'ai chi* appears, which will later be applied to the yin-yang symbol and to the martial art of that name in the

fourteenth century. The phrase is usually translated as the "Great Ultimate" or "Supreme Ultimate." It appears here as the source out of which the Ten Thousand Things are generated, since the eight trigrams—made of every three-line combination of yang (solid) and yin (broken) lines and associated with eight natural elements—will further combine with one another to produce sixty-four hexagrams as images of all possible conditions under heaven.

As the West was later inquiring into units of matter, developing the periodic table of elements, Chinese sages were investigating units of change, of process, which they represented through the system of 64 hexagrams. Something fundamental was being expressed through this deceptively simple description of the "production of all things." In our own recent scientific discoveries we have found similar foundational numbers. What is called the Computational Basis in computing consists of eight three-bit strings, which can be written as 000, 001, 010, 011, 100, 101, 110, 111 (MC². Quantum Computing, 2009). Replacing 1 and 0 with a solid and broken line, these correspond to the eight trigrams. Sixty-four is also the number of possible ribonucleotide triplets (codons) in DNA (Wagner, 1994). The systems cannot be conflated, but there is more than mysterious chance at work here: sage scientists past and present are peering into the very heart of nature's operations and discovering there an elegant numerical order.

In the early part of the Sung dynasty (960-1279 CE), a philosophical renaissance occurred during which Confucian philosophers, challenged by the widespread influence of Buddhism in Chinese culture, turned back to this text as a source for their own Neo-Confucian philosophy. One of the greatest of these, Shao Yung (1011-1077), would mine the passage above from the *Ta Chuan*

---

[4] As noted, contemporary scholarship places them at a later date, but the origin of the ideas is no doubt from an earlier time and shows Confucian influence.
[5] In his classes on the *I Ching*, Yi Wu notes that the more accurate translation of *ming* is destiny, rather than fate, as it often appears in the Wilhelm/Baynes translation.

for both its mathematical, cosmological and spiritual implications. Here is his version of it:

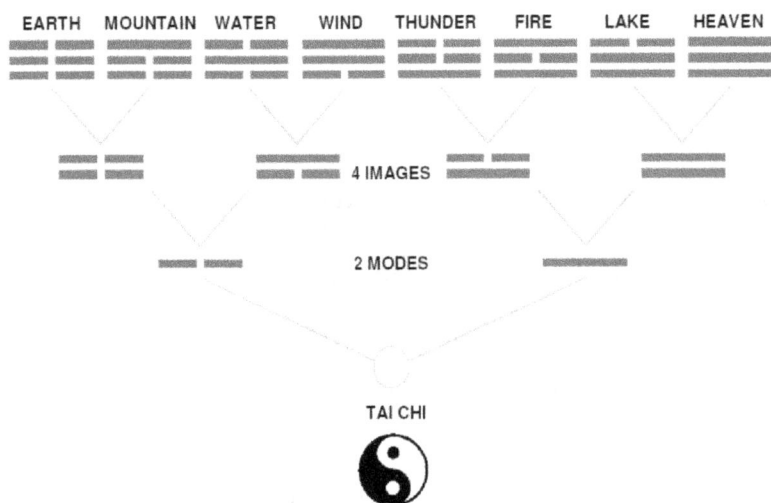

EARTH MOUNTAIN WATER WIND THUNDER FIRE LAKE HEAVEN

4 IMAGES

2 MODES

TAI CHI

T'ai-chi having divided, the two modes were established. Yang rose and interacted with yin. Yin descended and interacted with yang. [Thus] four images were born.

Thereupon the eight trigrams were realized (completed). The eight trigrams interacted, and afterwards the ten-thousand things were born therein (quoted in Smith et. al., 1990, p.111)

It was natural that Shao Yung, as a mathematician, would want to graphically represent this scheme. He followed the ancient format of the *I Ching* using the solid line to represent yang (associated with light, warmth, firmness, action) and a broken line to represent yin (associated with darkness, cold, yielding, stillness). The graphic above illustrates the beginning of the process he used to arrive at his own arrangement of the hexagrams based on chapter I.11 of the *Ta Chuan*.[6]

Shao Yung continued the process implied in the *Ta Chuan* chapter by doubling each trigram, adding to one, a yin line and to the other, a yang line, thus producing two new figures from each one or sixteen in total. He repeated the process to produce thirty-two, and again to produce the sixty-four hexagrams of the *I Ching* system.[7] He then took another remarkable step and wound the

long line of sixty-four hexagrams into a circle, which is at once mathematically elegant and philosophically profound.

He was faithful to the principle of the *Ta Chuan* that the entire process was a unity and that there was but a single hexagram of "six empty spaces" through which the changing interactions of yin and yang moved. He likens it to a tree:

Ten divides and makes one hundred. One hundred divides and makes a thousand. A thousand divides and makes ten-thousand. It is like the way a root has a trunk, a trunk has branches, a branch has leaves. Unite them and it makes one. Spread them out and they make ten-thousand. (quoted in Smith et al., 1990, p.111)

Scientific American (1974) featured on its cover a circular image of the eight *I Ching* trigrams with the t'ai chi (yin-yang) symbol at its center. Martin Gardner's article recounted the story of the discovery by the seventeenth century German mathematician Gottfried Leibniz of Shao Yung's diagram, sent to him by a missionary friend from China. Leibniz had just invented the binary numbering system, from which he would create an early prototype of the computer. He was amazed to find that the figures in Shao Yung's diagram, if yin and yang lines were replaced by 1 and 0, proceeded in a perfect binary sequence from 0 to 63. Both Leibniz and Gardner were impressed by Shao Yung's discovery of a binary system in the eleventh century in China. But Shao Yung was simply giving schematic form to a system first put forth in the *Ta Chuan*. And the *Ta Chuan* was simply reflecting the acute observations

of nature itself by generations of sages like Yao.

Shao Yung would take yet a further step that brings the numerical process into clearer focus as the activity of an integral and integrating consciousness in which we see the subtle relationship between the One and its one-ing.

"The mind is the Supreme Ultimate."

"The Supreme Ultimate is one, unmoving. It produces two; when there are two is spirit (*shen*)."

"Spirit produces number, number images, when images, instruments."

"The Preceding Heaven [an early arrangement of the trigrams] learning is a training of the mind; therefore the charts are all developed from the centre. The innumerable transformations and activities are born in the mind (quoted in Graham, 1992, p.154).

One can hear the echo of Shao Yung in a description of twentieth century physicist David Bohm, reflecting on the new view of reality suggested by quantum mechanics:

The ultimate nature of physical reality is not a collection of separate objects (as it appears to us), but rather it is an undivided whole that is in perpetual dynamic flux. For Bohm, the insights of quantum mechanics and relativity theory point to a universe that is undivided and in which all parts "merge and unite in one totality." This undivided whole is not static but rather in a constant state of flow and change, a kind of invisible ether from which all things arise and into which all things eventually dissolve. Indeed, even mind and matter are united: "In this flow, mind and matter are not separate substances. Rather they are different aspects of one whole and unbroken movement. (Quoted in Keepin, 1993, Wholeness And The Holomovement section, para. 1).

Brian Swimme and Thomas Berry would expand on that idea in *The Universe Story* by saying that "the universe is not a collection of objects but a communion of subjects" and make it central to their vision of the Ecozoic Era (Swimme & Berry, 1994).

The movement of Tao as Opening and Closing

Another passage from the *Ta Chuan* is eerily resonant with descriptions of quantum reality and slippery electrons

---

[6] In this image, I have used the Tai Chi symbol to represent the Supreme Ultimate, although it was not yet in use at the time of the *Ta Chuan*.
[7] It should be noted that the resulting order of the hexagrams is not the same as the received order we have today, which appears to be intentionally random to support the oracular use of the work.

that go into and out of phenomena as waves and particles. In chapter I.5, we find a line that would literally be translated as "One yin, one yang is called Tao." This can also be read as "one-ing through yin and one-ing through yang is called Tao," expressing the integrative movement of energy into units of all kinds fashioned through the two hands of yin and yang. Wilhelm (1977) catches the subtle sense of this in his translation of this line: "That which lets now the dark, now the light appear is Tao" (p.297). The line is pointing to Tao functioning (through one-ing) as the gateway between the manifest and unmanifest, the phenomenal and what is before phenomena. (The characters for opening and closing both contain the element for "gate.") It also underlines the fact that yin and yang are not a duality but two modes of the single unfolding.

A parallel statement in the same chapter, literally translated "One opening, one closing is called change [*pien*]," tells us that Tao in the phenomenal world is the course of change itself. The *I* (*Yi*) of the *I Ching* is a word that also means "change"—(its character was thought to have originally depicted a chameleon)—but never appears in the work with that meaning. How could this be so? The answer may be that this *yi* is the great one-ing of the universe through Tao in action as a single movement (like a wave) of change that is the unfolding of the universe through time, like Bohm's "undivided whole in a constant state of flow and change." (Quoted in Keepin, 1993, Wholeness And The Holomovement section, para. 1). *I/Yi*, as "change," moves both through a creative or light force which "opens" (opens the "gate" from the un-manifest) and through a complementary dark force which "closes," enters embodiment or expression in form. Even the word "force" here is inaccurate because it is not two dynamics in the sense that modern physics defines "force," but two modes of a single creative process.

The line that follows this would literally be translated "Coming and going without exhaustion is called interpenetration." The phrase, "coming and going without exhaustion" echoes the language of Hexagram 48, "The Well," which may have inspired this section of the Great

Commentary: "Coming or going the well is the well." (The well was traditionally located in the center of the town.) The Confucist commentary on this says "the well nourishes and is not exhausted" (Wilhelm, 1977, p.630), underlining the importance of the central position: The center is "inexhaustible" because it is the threshold of Tao. Interpenetration is imaged here as a passage—a "coming and going"—through the doorway or gateway. Hence nameless, formless mystery, as it is described by Lao Tzu, "penetrates" into a phenomenal world. There is a reverse process which is the rare capacity of the sage who penetrates into what is hidden, specifically through the gateway of the *I Ching*, which was used to reveal the "subtle beginnings" (also translated as "the minutest springs") of things. In the words of the *Ta Chuan*: "The operations forming the *I* are the method by which the sages searched out exhaustively what was deep, and investigated the minutest springs of things" (Legge, 1998b, *Ta Chuan*, section 1).The sage thus becomes the door at the threshold of Tao.

The idea that coming into and going out of visibility and phenomena is inexhaustible is coupled here with the idea of interpenetration (*t'ung*). The implication is that through *t'ung*, a unity is established among the ten thousand things as All Things Under Heaven. *T'ung* in this sense has a direct relation to Tao and also to "one" (*i/yi*), as Tao in action through "one-ing." "One-ing" comes about naturally as both an opening and a closing process. It "opens" into various entities through penetration and yet never invades or disintegrates their nature but encloses them in a higher order of unity that is not a fixed permanence or even unchanging form but the "way" of the cosmos as Tao itself.

The Production of Images

Following the generation of the two modes out of the Supreme Ultimate, the two modes give rise to the four images. The next line of chapter I.11 of the *Ta Chuan* gives an explicit definition of image. "Seen (manifest) it is called image." It is as if the friction of the two modes interacting produces light and that allows the first images to be seen. What is brilliant about what we would call this ancient Chinese cosmology is that it includes the Unmanifest and the darkness of potential energy as the precursor to light and manifestation. It recognizes the ineffability of all that precedes the manifest world and its spirit-like qual-

The I Ching takes up the moment when the first things are seen as images because it is at this point that they can be used and become part of the great field of action that is the human enterprise.

ity. The *I Ching* takes up the moment when the first things are seen as images because it is at this point that they can be used and become part of the great field of action that is the human enterprise.

If we are looking narrowly for a source of material things, we are confined to "the building blocks of matter." If we are looking instead for a source of consciousness or cosmic imagination that forms out of the simplest archetypes or patterns, then we are looking for images or something close to what contemporary chaos and complexity scientists would call fractals. In the *I Ching*, the first images are forms of contrast (yin and yang), which interact and, through repetition into patterns, are observed as the four seasons or four phases of light, (also of degrees of heat and movement), representing the first distribution of energy in time and space. This is an event akin to an awakening of consciousness, even to the diurnal awakening out of darkness and sleep.

Every phase of manifestation—from its primal beginning to the everyday affairs of humans—is encompassed by the *I Ching*. Thus the first role of the *I Ching* sages was to examine as extensively as possible even the seemingly incoherent and "confused diversities" and find in them a coherence; "a pattern which connects" to use biologist Gregory Bateson's (1979) phrase. In this task, they are the self-organizing power at the "center" between Heaven and Earth, specifically denoted by the central (third and fourth) lines of the hexagram. This is the realm of the Human, which includes one line of the lower (Earth) trigram and one line of the upper (Heaven) trigram as one of the three "ultimates" or powers of creative process. The Shuo Kua (Eighth Wing) tells us that in the realm of Heaven, yin and yang manifest as the dark and the light; on Earth, as the firm and the yielding; in the realm of

mentary on the Image of Hexagram 3, "Difficulty at the Beginning." (p.264).

> Clouds and thunder are represented by definite decorative lines; this means that in the chaos of difficulty at the beginning, order is already implicit. So too the superior man has to arrange and organize the inchoate profusion of such times of beginning. (Wilhelm, 1977, p.17)

The Confucist commentary on Hexagram 38, which Wilhelm translates as "Opposition," could be a summary of these lines of the *Ta Chuan*. Here is Legge's translation.

> Heaven and Earth are separate and apart, but the work which they do is the same. Male and female are separate and apart, but with a common will they seek the same object. There is diversity between the myriad classes of beings, but there is

old of Tai Chi. This was usually achieved through a process of purification, meditation, and the capacity to discern the "subtle beginnings" of things in phenomena, both in inward experience and outward circumstance. As the human mind touches the One at the source, it shifts from what later Taoist sages called the Wandering Mind to the Shining Mind or the Mind of Tao. This is another expression of putting oneself in accord with Tao.

The influence of the work on later Taoist thought can be seen in a thirteenth century work by Li Tao-ch'un, *The Book of Balance and Harmony*. "Open non-reification" points to a condition of the mind in a state of opening (not "closing" on individual objects of thought) at the threshold of Tai Chi.

> The Tao is fundamentally utterly open;
> Open nonreification produces energy,
> One energy divides into two modes:
> The one above, clear, is called heaven;
> The one below, opaque, is called earth.
> When the heart is nurtured by openness,
> It thereby becomes still;
> When energy is nurtured by openness, it
>   thereby circulates.
> When the human mind is calm and quiet,
> Like the north star not shifting,
> The spirit is most open and aware.
> For one who sees this
> The celestial Tao is within oneself.
> (Quoted in Cleary, 1989, p.91)

Throughout the I Ching, the 'superior person' is the one who cultivates an integrative consciousness, derived from an awareness of the essential Oneness out of which the limitless creative variety arises.

the Human, as love [humane feeling] and rectitude (Wilhelm, 1977, p.264). In contemporary language, we would say that Heaven-and-Earth (the Cosmos) brings forth the human as a conscious, integrated reflection of itself. Only the human can celebrate and "sing praises to the Creator" in countless different ways through the manifold gifts of human consciousness, including the capacity for reverence and gratitude that the creation inspires.

One-ing as Gathering Together

Throughout the *I Ching*, the "superior person" is the one who cultivates an integrative consciousness, derived from an awareness of the essential Oneness out of which the limitless creative variety arises. Such a person exercises the capacity for bringing order out of chaos through his or her own integrity, "just as one sorts out silk threads from a knotted tangle and binds them into skeins." Wilhelm describes it in his com-

an analogy between their several operations. Great indeed are the phenomena and the results of this condition of disunion and separation. (quoted in Sung, 1969, p.164)

The text counsels further that "in order to find one's place in the infinity of being, one must be able both to separate and to unite" (Wilhelm & Baynes, 1983, p.17).

One-ing as discernment of the invisible unity

> Variation and transformation originally came from one, but through interaction, arrive at the infinite *li* [patterns; principles]. When people observe this, they think there are infinite differences. But when the sages observe this, wherever they go, it is one. (Su Shih quoted in Smith et al.,1990, p.74)

Through what is sometimes called "the reverse path of sages," a sage has the capacity to move in the reverse direction from the multiplicity of things to return to their source beyond the thresh-

old of Tai Chi. (continued implied)

The Virtue of the Way – Tao Te

This singleness of the way (tao) of Heaven and Earth, this one-ing in action, is the meaning of *te* in the *Tao Te Ching*. (Ching means "classic book.") The word is usually translated as "virtue," or "power." The character is made up of the element for heart-mind (inseparable in Chinese) and above that, the single line meaning "one" or "one-ing." Above that, are "ten eyes," meaning "ten eyes have seen it and called it good." To the left is the abbreviated character for "walking." I like to give a literal translation for this character as "Single-heartedness (-mind-edness) in action before the world." Just as the singleness of Tao as Oneness unfolds in the world as the Way of Heaven and Earth—without doubleness—so too does the human (the third power arising in their midst) unfold an integrity of being into a creative integrity of action.

When the human increasingly senses

his or her natural course (tao) within the larger course (Tao) as the Way of Heaven-and-Earth—whether experienced in the inmost being or the external world—*te* arises as this force of virtue which is both the singular flowering of the individual nature and the quality of the universe coming into being. Virtue thus becomes the inner force of One (Tao) in action; the quality of one-ing. This is the meaning of a phrase that appears in the *I Ching* and the *Great Learning*—*ming ming te*: to manifest bright virtue. *Ming*, the "light, clarity" of natural inner virtue is brought outward to shine forth as the second *ming*. Something given by nature is consciously and deliberately cultivated and lived out as the whole or integral person.

Joel Primack and Nancy Abrams, while presenting a detailed picture of our latest view of the universe as translated to us by the most advanced technologies of optics and computers, pose the question—"How many people recognize the possibility of a sacred relationship between the way the expanding universe operates and the way human beings ought to behave?" (Abrams & Primack, 2001, p.1769). They are pointing to a rift between our sophisticated knowledge of how the universe works and the lack of a commensurate wisdom about how the human behaves in that universe.

Correlations are beginning to appear between the language and images of science, psychology and the classics of philosophical and religious traditions. What is still missing is a more directly articulated view of the role of human consciousness and conduct in these processes. The new sciences of chaos and complexity theory show that throughout the living world, chaos is constantly being transformed into order. Physicist Ilya Prigogine, for example, introduced the theory of dissipative structures.

> In classical thermodynamics the dissipation of energy in heat transfer, friction, and the like was always associated with waste. Prigogine's concept of a dissipative structure introduced a radical change in this view showing that in open systems dissipation becomes a source for order.... When the flow of energy and matter through them increases, they may go through new instabilities and transform themselves into new structures of increased complexity. (Quoted in Capra, 1996, p.88)

Prigogine identifies a bifurcation point that is a threshold of stability at which the dissipative structure may either break down or break through to one of several new states of order. For Jung, this language would describe the key point in the psychic life of an individual where a peak of tension can lead to psychosis and breakdown or regeneration and rebirth as the psyche breaks through to a higher level of awareness and creativity.

According to Prigogine, what happens at the bifurcation point is dependent on the system's previous history. "Living structure is always a record of previous development" (quoted in Capra, 1996, p.191).

> At the bifurcation point, the dissipative structure also shows an extraordinary sensitivity to small fluctuations in its environment. A tiny random fluctuation, often called "noise," can induce the choice of path. Since all living systems exist in continually fluctuating environments, and since we can never know which fluctuation will occur at the bifurcation point at the "right" moment, we can never predict the path of the system. (Capra, 1996, p. 191)

We can, however, imagine a correlation between sensitivity to small fluctuations in the environment and the sage's ability to perceive subtle beginnings, the seeds of things, and to seize the right moment for the right response. This is what we have defined earlier as the self-organizing power of the human positioned at the center between heaven and earth. The scientist has the extraordinarily amplified capacity, aided by sophisticated techniques and instruments, to enter these processes at a minute level. Yet this is only one human capacity, which should itself be guided by a yet more subtle capacity to determine what direction, in any given moment, is most healthy, beautiful or furthering for the system of which one is a part; i.e. through personal integrity, (reflecting the cosmic or divine integrity), putting oneself in accord with what is right. From the perspective of the *Ta Chuan*, "what is right" is that which accords with the grain of the universe as the one-ing of its creative advance; the will or "mandate" of heaven as it is described in this and other classical texts.

The religious context enters here with its affirmation that the human can mysteriously experience the cosmic whole; can be moved by the Mind of Tao. The Hindu sage is moved by Brahman and the Bodhisattva by the Buddha-mind toward an experience of the "whole" that includes an active compassion for all created beings as its most immediate manifestation. In Jesus' last prayer before surrendering to the events of the crucifixion, he prays, "May they all be one: as thou, Father, art in me, and I in thee, so also may they be in us. The glory which thou gavest me I have given to them, that they may be one, as we are one; I in them and thou in me, may they be perfectly one" (John 17:21-23; [NEB, p.184]). Jesus at this moment is sacrificing himself to the higher value which is thus kept alive in the civilization with his death and resurrection.

Lastly, our own Western worldview, which has brought us to a high point of individual consciousness and material development, is now going through a dramatic transformation in which the human, seen from a post-Copernican scientific perspective as "outside" of nature, is restored to a central position "between Heaven and Earth," essentially embedded within the functioning of a holistic universe. Richard Tarnas speaks eloquently of this shift.

> We are not ultimately separated from the world, projecting our structures and meanings onto an otherwise meaningless world. Rather, we are an organ of the universe's self-revelation. We are beginning to see that we play a crucial role in the universe's unfolding by our own cognitive processes

This singleness of the way (tao) of Heaven and Earth, this one-ing in action, is the meaning of te in the Tao Te Ching.

and choices, tied to our own psychological development. Our own inner work—our moral awareness and responsibility. (Tarnas, 1998, A Period of Transition section, para. 3)

The most developed human, entering into the highest levels of conscious relationship to the world, is constantly at this threshold of decision with the possibility of moving toward increasing order, creativity, harmony, integrity and love, even as the least developed human impulse leads toward violence and disorder. If the human is indeed embedded in a system of interrelationships, then he or she shares in the "history" of each moment not only personally but collectively. From the perspective of the global mind, our collective historical moment brings us to a bifurcation point of great magnitude and import that calls upon the inner integrity of each of us—our "inner sage"—to respond with care to the subtlest beginnings of what is to come.

References

Abrams, N. E., & Primack, J. R. (2001). Cosmology and 21st-century culture. *Science, 293*(5536), 1769-1770.

Bateson, G. (1979). *Mind and nature: A necessary unity*. New York: E.P. Dutton.

Capra, F. (1996). *The web of life*. New York: Doubleday.

Cheng, C.-Y. & Bunnin, N. (Eds.). (2002). *Contemporary Chinese philosophy*. Oxford: Blackwell.

Cleary, T. (Trans.). (1989). *The book of balance and harmony*. New York: North Point Press.

Gardner, M. (1974). Mathematical games: The combinatorial basis of the *I Ching*, the Chinese book of divination and wisdom. *Scientific American*. 230(1), 109-13.

Graham, A.C. (1992). *Two Chinese philosophers*. La Salle, IL: Open Court Press.

Keepin, W. (1993). Lifework of David Bohm—River of truth. *ReVision, 26*(2). Retrieved April 28, 2002 from http://www.vision.net.au/~apaterson/science/david_bohm.htm

Legge, J. (Trans.). (1971). *Confucius: Confucian analects, the great learning and the doctrine of the mean*. New York: Dover.

Legge, J. (Trans.). (1998a). *Shu jing (Shu ching)*, vol.3. First published 1960. Reprint, Hong Kong: Hong Kong University Press. Retrieved April 15, 2002 from http://www.chinapage.org/confucius/shujing-e.html

Legge, J. (Trans.). (1998b). *Ta chuan, The great appendix*. First published in 1899. Retrieved April 19, 2001 from http://www.mindsports.net/I_Ching_Connexion/Appendices/ compiled by Christian Freeling.

Lynn, R. J. (Trans.). (1994). *The classic of changes: A new translation of the* I Ching *as interpreted by Wang Bi*. New York: Columbia University Press.

MC². Quantum Computing. (2009). Retrieved September 2, 2009, from http://mc2.gulfpixels.com/?p=411

New English Bible, New Testament [NEB]. (1961). Oxford & Cambridge: Oxford University Press and Cambridge University Press.

Smith Jr., K., Bol, P.K., Adler, J.A., & Wyatt, D.J. (1990). *Sung dynasty uses of the* I Ching. Princeton: Princeton University Press.

Sung, Z.D. (1969). *The text of Yi King (and its appendices)*. Chinese original with English translation [based on the James Legge translation]. New York: Paragon.

Swimme, B. and Berry, T. (1994). *The universe story*. San Francisco: HarperOne.

Tarnas, R. (1998). "The Great Inititation." *Noetic Sciences Review (47)*. Retrieved July 6, 2009 from http://www.noetic.org/publications/review/issue47/main.cfm?page=r47_Tarnas.html

Wagner, R. P. (1994). Understanding inheritance: An introduction to classical and molecular genetics. In N. G. Cooper (Ed.), *The human genome project*, ed (pp. 1 -67). Sausalito, CA: University Science Books.

Wieger, L. (1965). *Chinese characters: Their origin, etymology, history, classification and signification*. New York: Dover.

Wilhelm, R. & Baynes, C. (Trans.). (1977). *The I Ching or book of changes (3rd ed.)*. Princeton: Princeton University Press.

Wu, Y. (Ed. & Trans.). (1989). *The book of Lao Tzu (The tao te ching)*. San Francisco: Great Learning Publishing.

Wu, Y. (1990). *Chinese philosophical terms*. San Francisco: Great Learning Publishing.

Wu, Y. (Ed. & Trans.). (1998). *I Ching: The book of changes and virtues*. San Francisco, CA: Great Learning Publishing.

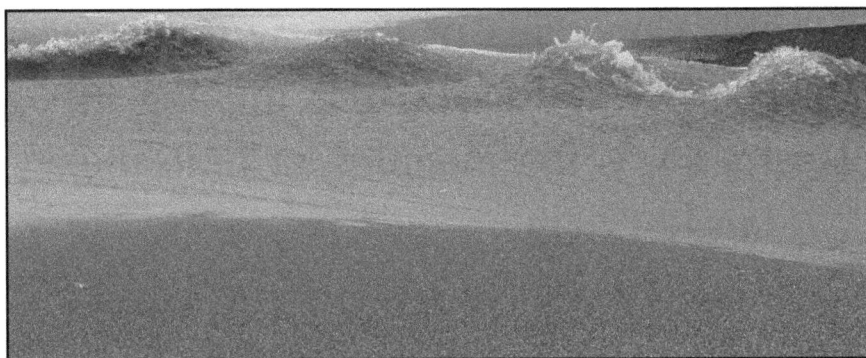

Photo: Jürgen Werner Kremer

Millennium Poem

This gently being on the earth
will last forever.
This gently going to and fro
will go on forever.
When someone asks,
"What is the millennium?"
Tell them it is this:

gently being on the earth,
gently going to and fro.
This is love eternal,
this,
simply this.

Michael Sheffield

The Role of the Astrological Symbol System in Understanding the Process of Evolutionary Growth

Armand M. Diaz

Astrology is a symbolic system that began thousands of years ago and that has continued to develop, deepen, and expand to the present day. The place of astrology within the community of people who seriously consider the evolution of consciousness, however, is less than secure. This article will offer a perspective on astrology that might be useful to help incorporate astrological ideas into the larger framework of evolutionary thought. The first part of the article is intended to show that astrology is potentially compatible with perspectives coming from the new paradigm sciences — a group of related approaches that challenge the dominant Newtonian-Cartesian world view— which is emerging in the physical and social sciences. For this, I will borrow an idea from chaos theory in order to show that astrological influences need not be construed as deterministic, and I will use Jung's (1973) concept of synchronicity to demonstrate that astrology need not have a causal basis or be construed in terms of mechanistic determinism. The second part of the article will deal

Armand M. Diaz is a graduate student in the Transformative Studies program at the California Institute of Integral Studies. Contact: Armand M. Diaz, 3327 201st Street, Bayside, NY 11361. Email: Zitofon@Verizon.net.

with the question of astrology's potential usefulness, for, after all, even if astrology has validity the community of transpersonally oriented theorists, practitioners, and researchers would still want to know how it can contribute to personal and collective consciousness evolution. This second part will therefore build upon the first, incorporating a model of the development of consciousness in order to explore one of the thorniest problems in astrology (and divination in general): prediction.

Archetypes and Attractors

Although this article is not intended as a vehicle to refute the criticisms of astrology, it is necessary to say something about the way astrologers view their own discipline in contrast to the way it is seen by critics. While some astrologers do believe that there is some physical explanation for astrology, the majority see astrology not in terms of efficient or material *causation*, but rather as a *meaningful correspondence* between the movements in the sky and what happens here on Earth, as suggested by the Hermetic saying, "as above, so below," which has been the unofficial philosophical axiom of astrology for centuries. In his book *Cosmos and Psyche*, Richard Tarnas (2006), the most recent and probably

the most eloquent exponent of this view, elaborates on how astrology is not a system of physical influences but of synchronicities relating the meaningful coincidence of planetary configurations with the underlying archetypal themes of human experience.

This movement away from causal determinism is an important step in seeing astrology as astrologers do, because when astrology is freed from the constraints of trying to be understood as a physical science, it can be judged and evaluated on its own terms. For example, rather than the planet Mars *causing* violence or injury through some kind of emitted force or influence, one should recognize that Mars coincides or corresponds with a particular set of symbols signifying assertiveness, aggression, drive, and so on. Furthermore, while astrology cannot produce reliable, replicable results on the physical level when applied to the concrete particulars of human experience, it can demonstrate *symbolic cohesiveness* on the level of meaning.

Additional complexity is added when we bear in mind that symbols or the archetypes behind these symbols are inherently productive and creative. That is, there appears to be no end to the number of ways that the symbol can manifest, because it does not merely stand for a fixed set of possible items, but for an infinite variety of possible manifestations that all share the same basic meaning. As Tarnas, (2006, p. 67) has said, the nature of astrological correlations is "the multidimensional and multivalent nature of archetypes — their formal coherence and consistency that could give rise to a plurality of meaning and possible manifestation." Add that astrological symbols always manifest in combination (e.g., the *planet* Mars in the *sign* of Gemini and in relationship to other planets with their own set of symbolic meanings), and a rich variety of potential meanings emerges.

Many astrologers, including Tarnas (2006), speak of astrological symbols as archetypes. This makes sense in that they appear to be somehow related to the basic structures of both consciousness within and the world without, as in the Platonic and late-Jungian understanding of archetypes. The Moon, for example, represents our emotional nature and responsiveness, and is also related to the Great Mother archetype, the nurturer, to the archetypal principle that brings things into being and cares for them. The archetypal approach to astrology recognizes that specific cosmic events (i.e., the shifting pattern of planetary alignments) will have meaningful correspondence in the lives of people (and of nations, communities, and sometimes the planet as a whole), although Tarnas (2006) stresses that the exact way the symbols will play themselves out in the specifics of human experience cannot be known, only the archetypal meaning of these symbols.

An archetypal approach to astrology allows for a kind of symbolic playfulness on the part of the cosmos, a freedom from the constraints of rigid determinism. Indeed, the language of archetypes is often presented in terms of the fluid symbolism of mythology, and this can be an excellent approach to understanding astrological factors. As a practicing astrologer, I value the

> **Cases are seen in which the predictions of the horoscope fulfill themselves with great accuracy up to a certain age, then apply no more. This often happens when the subject turns away from the ordinary to the spiritual life. If the turn is very radical, the cessation of predictability may be immediate; otherwise certain results may still last on for a time, but there is no longer the same inevitability. This would seem to show that there is or can be a higher power or higher plane or higher source of spiritual destiny which can, if its hour has come, override the lower power, lower plane or lower source of vital and material fate of which the stars are indicators.**
>
> **(Sri Aurobindo, 1970, p.468)**

rich symbolic value of mythology, as it often provides a key to unlock meanings deeper than a more rational explanation can provide. However, because astrology is often associated exclusively with a mythic world view (Wilber, 1997), the use of mythic images has the unfortunate consequence of reinforcing the idea that astrology itself is mired in a lower, pre-rational level of thinking. In addition to its close association with myth, however, the workings of astrology can, I believe also be approached by using certain concepts and theories from the new sciences. When understood in these terms, astrology might become more palatable to the modern, scientifically informed mind.

Borrowing from chaos theory, I want to suggest that an astrological factor (such as the archetypal principle associated with the planet Mars) can act as a kind of *strange attractor*, which Briggs and Peat (1999) describe as a pattern of

behavior in which "the system's behavior is unpredictable and non-mechanical. Because the system is open to its external environment, it is capable of many nuances of movement" (Briggs & Peat, 1999, p. 64). In astrology, being open to the external environment means that any particular astrological factor will have to be seen in light of all the other factors in an astrological chart and, more generally, in the context of all the many other factors that shape human experience, to greater or lesser degrees. For example, the planet Mars is always associated with a particular type of energy which is inflected by the sign the planet is in and by its relationship to the other planetary bodies, all of which are associated with distinct archetypal meanings of their own which may dynamically interact with each other. That is, we could say that the archetypal patterns associated with Mars are, like those of a strange attractor, reliably recognizable, but that these are always open to modification by a variety of astrological factors. Additionally, a person or thing here on Earth will express the symbolic meaning of Mars in light of many other non-astrological factors such as species, gender, culture, and all of the other myriad conditions to which beings are subject. This article will explore one of these factors: the development of consciousness.

Astrology in Light of a Model of the Development of Consciousness

Wade's (1996) holonomic model of the development of consciousness is a comprehensive and inclusive approach to development, coordinating the work of previous investigators and theoreticians, such as Piaget (1976) and Maslow (1982), and recognizing the correlations between ego development, neurological function, and social organization. Wade identifies a number of developmental levels: reactive, naive, egocentric, conformist, achievement, affiliative, authentic, transcendent, and unity (Wade, 1996). Prenatal, perinatal, and after-death states are also included in her scheme. The developmental levels are experienced sequentially, moving from reactive to unity consciousness in order; and each successive stage is deemed to

Astrology is not a system of physical influences but of synchronicities relating the meaningful coincidence of planetary configurations with the underlying archetypal themes of human experience.

be more complex than previous stages, while also dependent on the earlier stage of consciousness (Wade, 1996, p. 21). That is, one cannot skip stages, although achievement and affiliate consciousness are at the same level of cognitive complexity and represent two paths to authentic consciousness in this model. While each successive level represents increased awareness and greater flexibility of response over previous levels, it is also true that each level contains a pathological side, although the authentic level and above have not been studied in this regard; i.e., a relative improvement in moving from one level to the next does not imply that pitfalls of various kinds do not remain, which, we will see, is significant as we apply this model to astrology.

According to many transpersonal theorists, how any phenomenon is viewed is partly determined by the developmental level of consciousness of the individual. For example, in *Integral Spirituality* Wilber (2006) argues that religious texts and symbols will be interpreted differently by individuals at different developmental levels. This approach can also be applied to astrology, resulting in a variety of forms of astrology which vary not only by technique, but also by the level of consciousness they reflect. It may be

true that astrology entered the stream of human thought when the consciousness of most people probably did not stray beyond the egocentric level, which corresponds to what Wilber refers to as the magic-mythic world view (Wilber, 2000, p. 207). However, there is really no reason to assume that astrology per se belongs to those levels. The history of astrology is long and complex, and even if we limit the focus to that which was to become modern Western astrology we would need to look across many cultures through thousands of years to get a fuller picture of the many different levels of insight and knowledge that have contributed to shaping this movement. However, it is essentially at the mythic level that astrology entered the picture, as demonstrated by the zodiac images, and the naming of the planets after gods and goddesses, and it is to this level that it astrology generally consigned by critics (Wilber, 1997).

That consignment would be appropriate if astrological thinking had not advanced beyond the mythic level. But a

been reflective of the magical level that dominated at the time, but this gave way to the mythic level, and later to the subsequent levels, so that today astrologers, or at least *some* astrologers, can practice from higher levels, as demonstrated by the so-called "psychological astrology" that became popular in the 1970's, resonating with Wade's affiliative level of consciousness.

A review of the current literature would probably show that astrologers today represent a variety of developmental levels in their approach to astrology. To be sure, there are still some astrologers who make fatalistic pronouncements and concrete predictions, but there are also calls for greater self-reflective understanding of ourselves, and an acknowledgement of the evolutionary potential we possess (e.g., Forrest, 2008). Therefore, while it would be unfair to say that all astrology is mired in a lower level of development, it is not unreasonable to say that some forms of astrology are indeed quite limited, as long as we recognize that some forms of astrology are also in concert with more advanced developmental levels.

I say that astrology may be "in concert with" higher developmental levels because I don't want to overemphasize astrology and its role — a mistake that I think is common among astrologers who have ventured beyond the conformist

An archetypal approach to astrology allows for a kind of symbolic playfulness on the part of the cosmos, a freedom from the constraints of rigid determinism.

reading of astrological literature over the centuries shows that this is clearly not the case. Astrologers have been neither undeveloped laggards among their contemporaries, nor, in my opinion, are they developed in *some* respects but trapped at a lower level in terms of their assumptions about and practice of astrology. Rather, their thinking about astrology has generally followed the path of the culture in which they live. The fatalistic omens of the earliest astrology may have

level. While I don't intend to criticize the idea that astrology itself contains a developmental scheme out of hand, I think that astrology is better understood as a sort of weather report for what types of archetypal situations and themes will emerge at each level. In other words, astrology does not describe the developmental process itself, but the "cross-winds" — the archetypal dynamics and themes — that "blow" at all levels. If these archetypal factors have any role at all in promoting psychological or spiritual development, I think that it is through their forms of expression within situations *at the level a person is already on* which may create a stress, life opportunity, or adaptive challenge that can only be resolved successfully by moving to a higher level. The archetypal principles associated with the planets seem to give rise to experiences that can impel and inform our growth. But then again, there is no guarantee that we will successfully resolve the various challenges we encounter, and regression or stagnation are also possibilities. A great deal depends on the conscious attitude and degree of awareness of the person involved.

Astrological Prediction and Consciousness Development

I would like to demonstrate how these "cross-winds" operate by using a controversial example, that of astrological prediction. Prediction is perhaps the most problematic aspect of astrology because it seems to point to specific, fatalistic outcomes. Wilber (1997) predicts that astrology will "fail tests of accuracy when compared to methodologies derived from the mental (and higher) stages of development" (p. 328), thus suggesting that astrological prediction may work for the lower levels of development but will not work above the mythic level. *Accuracy*, however, is itself a criterion that can be variously defined. If astrological predictions are understood only as specific outcomes that manifest in the physical world (i.e., as events in a person's life), there is probably little reason to pay much attention to them. This can be resolved in part by moving to an archetypal model, as Richard Tarnas (2006) suggests. In this view, we can know what types of situations will emerge under given conditions, but cannot know what specific form they take, in that they are archetypally and not concretely predictive. Additionally, as we are concerned primarily with the meaning of the symbol rather than events, we cannot rule out the possibility that the manifestation of

a given set of astrological factors will take the form of an internal experience, such as a period of exhilaration, enthusiasm, or melancholy rather than as an external event. Thus, to adequately measure the accuracy of astrological predictions, we would have to account for a range of possible outcomes that manifest on the physical, social, psychological and perhaps even spiritual levels. One potential problem with the archetypal approach, however, is that the range of outcomes might become so inclusive as to effectively trivialize the value of prediction altogether. After all, there is probably only minimal value in knowing that an individual may have some broadly defined type of experience, and that value will be smaller still in the context of evolutionary development.

Viewing astrology, and specifically astrological prediction, through the lens

The astrological factors manifest in a variety of different ways according to developmental levels of consciousness."

of a developmental model of consciousness such as Wade's (1996), however, offers a way to both narrow the possible manifestations and to place such experiences within the context of the potential for evolutionary change.

To apply Wade's model to astrology, we first need to consider how astrological prediction works. The predictive techniques used in astrology are many and complex, and we don't need to consider all their variations. I will use the example of what is known as a *transit*, in relationship to an individual birth chart, which is essentially a map of the positions of the various planetary bodies and other significant astronomical points at the time of a person's birth. Which bodies are included in an astrological chart, and which zodiac is used as a frame of reference varies among astrologers, but each of the astrological factors used symbolizes a particular kind of energy, apparent in all our lives, as, for example, Venus represents relationship and love, and also beauty and the aesthetic sense. The positions of the planetary bodies at birth remains constant, as depicted in the

astrological birth chart, but of course the planets in real time continue to move in their orbits after birth. At various points in time, the moving planets will come into some kind of alignment with the positions of the planets, and other significant points, at birth. They may, for example, come to the same place in the zodiac that the body occupied at the time of birth, or may be 180 degrees across the zodiac from that position, and so on. These various relationships between moving bodies and the birth chart are called *transits* in astrology. For example, the planet Uranus, which is associated with a freedom-oriented and somewhat erratic and destabilizing energy, might transit the position of Venus in a person's birth chart. When this happens, over the course of a period of time, the dimensions of experience indicated by the *transited* planet (Venus) are affected in a way characteristic of meanings associated with the *transiting* planet (Uranus). In very broad terms, a person who is having such a transit might typically want to be free to pursue new and different types of relationships, or they might rebel against constraints in a present relationship, or they might be attracted to new forms of art, etc., as the Venusian principle (associated with beauty, romance, love, valuables, and so on) is affected by the archetypal principle of transiting Uranus for the duration of the transit.

Now, if we take a strictly archetypal view of astrology, we can predict the general flavor or thematic quality of the transit (e.g., changes in relationship) but we cannot say much about how the transit will actually manifest. For many reasons, especially those stemming from ethical considerations, I believe an acknowledgement of the archetypal, non-specific nature of astrology is the right approach to take when actually doing predictive work. Most astrologers would add that I have neglected to mention other factors, such as the sign and house positions of the planets involved, which help to narrow down the predictive range — a valid complaint against my simplified description of predictive work. Still, even after narrowing down possible manifestations of a transit, a fairly wide range of potential outcomes remain. This leads to the question: is there any way that we can gauge the impact of a transit on a person's life? I think the answer is a qualified yes, and the key is the level of consciousness development the person has attained in the affected area of her life.

The example I have been using, Uranus transiting Venus, is a relatively rare and powerful transit, occurring about every 19 to 21 years in a person's life, and lasting for between one and two years during each major transit. Among some astrologers, fairly dire predictions for this transit are not uncommon, including the break-up of a marriage or relationship, and other consequences of the urge for freedom at all costs. It is also true however, that many people go through this particular transit and emerge with their relationships intact, thus it seems the accuracy of astrological prediction is in question.

My work with clients has led me to an understanding of prediction, and indeed all of astrology, which is more complex than the concretely predictive approaches to astrology. In my experience, the astrological factors manifest in a variety of different ways according to developmental levels of consciousness. I should caution that I have neither the expertise nor the opportunity to conduct formal assessments of my client's development, and I base my tentative conclusions on the hints that have emerged in the course of consultations. The level of development each person attains may differ across different areas of life, which Wilber (2006) refers to as *developmental lines*, and they are roughly represented by different astrological symbols. Mercury, for example, relates to cognitive development, Venus to aesthetic development, and the moon to emotional development, although these generalities should not be taken to imply that astrological symbols can be easily grafted onto any other particular scheme. I have often observed that clients tend to consult astrologers on the issues where they are experiencing the most tension or dissatisfaction, perhaps pointing to dimensions of experience of lines of development that have not been sufficiently addressed and worked through

their lives. The discrepancy between a person's overall level of development and that of the particular area (or line) seems to throw the problematic area into a sharper relief.

Returning to my example of a transit of Uranus to Venus, it is my experience that those people whose relationships have remained at the conformist level (or below) will often have an experience that could be read out of the most fatalistic astrology text. During a powerful transit of Uranus to Venus, people at the conformist level will indeed often go through extremely stressful periods in their relationships. For example, those within partnerships sometimes experience divorce and separations, whether through their own doing or that of their partners. Obviously, since a large proportion of society may be assumed to be at this level, divorce is not the only outcome for married people having a Uranus-Venus at the conformist level, but whatever the ultimate result of the transit, it is usually acted out as a drama in the person's life. Another typical outcome, whether the person is single or in a relationship, is the experience of being attracted to someone who is not the person's usual "type." As belonging to a group and participating in the group's values are key to this level of consciousness (Wade, 1996), the disruptive quality

If a developmental approach were taken, it might be used to refine astrological prediction and to guide and inform the therapeutic process.

associated with Uranus rather frequently leads to a stress on the conformist's values, often resulting in the unconscious acting out of the Uranus-Venus drama. It is not atypical for a client in this position to be faced with the possibility of ending a marriage and expressing considerable distress because "we" don't believe in divorce. Similarly, being attracted to a potential partner from a different ethnic or religious group can be very stressful to the conformist perspective.

Beyond the conformist level, the achievement level is, in my experience, rather underrepresented among astrology clients, perhaps because people at this stage are more concerned with developing their personal power and see the use of astrology as capitulating to an external locus of control, or perhaps because astrology is incompatible with the logical, analytical mindset which Wade (1996) says characterizes this level. However, people at this level typically have a very different experience of the Uranus-Venus transit than those at the conformist stage. For example, relationships can go through stressful periods and sometimes do end in separations, but it is more likely at this level that the couples will go into therapy or seek some outside help, taking the proactive approach that Wade (1996) associates with achievement consciousness. Rather than simply reacting to the stresses on the relationship, people at this level seek to understand and gain some control over them.

At the affiliative level, the stresses upon a relationship might be viewed as an opportunity to consider how each partner is participating in the relationship, or to address the need for individual space to grow and develop as balanced against the needs of the partnership. Love and relationship are primary values and motivations at this level (Wade, 1996), and an approach which seeks to deepen understanding of the dynamics of partnership and a willingness to "love more" characterize the response to the Uranus-Venus transits at this level.

"What will happen?" is the most frequent question I hear from clients who appear to be at the conformist level or below in the particular area of life affected by an astrological transit, while at the achievement and affiliative levels the question becomes "what can I do?" At higher levels still (which, I have found, are also underrepresented among clients), the question seems to become, "what does it mean?" The experience will continue to reflect the archetypal nature of the transit, while the tendency to create concrete events in the form of life dramas diminishes. As an astrologer, I suspect that each entity with a birth chart, from a duck to a bodhisattva, will undergo an experience during a transit of Uranus to Venus that reflects the archetypal meaning associated with these planets. For the enlightened being, it is perhaps a passing thought on freedom and love that enters and leaves the mind without disturbing anything, but my guess is that some experience typical of the Uranus-Venus combination occurs just the same.

One hopeful possibility that should not be neglected when considering the developmental implications of transits is that the transit itself will create enough inner tension that awareness begins to grow and the client's perspective actually evolves to the next level, as opposing energies demand resolution by an adaptive shift to a higher level of psychological organization. It is also important to realize that while the outward, material manifestation of a transit may appear more benign at the higher developmental levels, this isn't necessarily the case, for often much of the stress is transferred to one's inner experience rather than being unconsciously acted out. One woman who doubted whether she would ever fall in love again because personal relationships seemed to her, at the time of a Uranus-Venus transit, to be too limiting (a rather abstract concern compared to, for example, the possibility of divorce) was experiencing a degree of anguish and sadness that could perhaps have rivaled some people's experience of the loss of their partnership. In other words, the "fated" appearance of the transit — the tendency for the transit to be experienced as events in a person's

life — diminishes as we go up the developmental chain, but the experience of challenge remains, and that challenge will generally be felt as very real, at least in the lower and middle ranges of development.

Wade (1996) points out that many of the theorists whose ideas she incorporated into her work stopped their model of development at their own developmental level, failing to see anything above that level. Astrologers, like teachers and therapists, often find themselves in the role of coaxing others towards their own developmental level, and as I am something of a laggard myself, things get very speculative for me above the middle ranges of the scheme.

Stanislav Grof (2009), working with Richard Tarnas, has observed that astrological transits often indicate the times at which individuals will have significant breakthroughs during experiential therapies. While the published accounts of this phenomenon are rare, there is a subset of psychotherapists who know and value astrology (whether or not they use it with their clients) and they also, anecdotally and informally, report that astrological transits can be helpful in predicting significant progress in therapy. However, it would clearly not make sense to say that an astrological transit by itself could predict either a significant breakthrough within a therapeutic context or an evolutionary advance in a person's life, as powerful astrological transits obviously outnumber epiphanies and quantum leaps in level of development, and do not usually manifest in this way. On the other hand, if a developmental approach were taken, it might be used to refine astrological prediction and to guide and inform the therapeutic process, giving insights into how particular astrological archetypal themes associ-

ated with different astrological symbols might manifest at different levels of consciousness.

References

Briggs, J., & Peat, F. D. (1999). *Seven life lessons of chaos: Timeless wisdom from the science of change.* New York: Harper Collins Publishers.

Forrest, S. (2008). *Yesterday's sky: Astrology and reincarnation.* Borrego Springs, CA: Seven Paws Press.

Grof, S. (2009). Holotropic research and archetypal astrology. *Archai: The Journal of Archetypal Cosmology, Issue 1.* Retrieved from www.archaijournal.org.

Jung, C.G. (1973). *Synchronicity: An acausal connecting principle.* (R.F.C. Hull, Trans.). Princeton, NJ: Princeton University Press.

Maslow, A. (1982). *Toward a psychology of being* (2nd. ed). New York: Van Nostrand Reinhold.

Piaget, J. (1976). *The Child and Reality.* New York: Penguin Books.

Sri Aurobindo. (1970). *Letters on Yoga.* Pondicherry, India: Sri Aurobindo Ashram Trust Publications.

Tarnas, R. (2006). *Cosmos and Psyche: Intimations of a new world view.* New York: Viking Press.

Wade, J. (1996). *Changes of Mind: A holonomic theory of the evolution of consciousness.* Albany: State University of New York Press.

Wilber, K. (1997). *The Eye of Spirit: An integral vision for a world gone slightly mad.* Boston: Shambhala Publications.

Wilber, K. (2000). *Integral psychology: Consciousness, spirit, psychology, therapy.* Boston: Shambhala Publications.

Wilber, K. (2006). *Integral Spirituality: A startling new role for religion in the modern and postmodern world.* Boston: Integral Books.

A Feminist Approach to Three Holy Women's Intuitive Processes

Exploring Aspects of Transcendental Phenomenology & Hermeneutics

Through a Correlation of the Researchers' Experiences

Ana Perez-Chisti

As you read these stories of three Holy Women who are dedicated to a life of devoted service and commitment to spiritual understanding, observe what moves you in your arising emotions. These Holy Women have much to offer us. They unite us in a common investigation of our reasons for being embodied. They offer a natural epistemological investigation into the way we can discover our intuition, our essence, and our true purpose. As a postmodern theorist who practices the inversion of mind-body dualisms by showing the body as a sound source of knowledge, I decided to apply a feminist approach to three narratives because preference could be given to women's felt-experiences as a

Ana Perez-Chisti, PhD, is faculty member and former Chair of the Global PhD Program at the Institute of Transpersonal Psychology in Palo Alto, California. She is a lineage holder and National Representative of the Sufi Movement International and is an ordained minister and senior teacher (Murshida). She has lectured internationally bringing the subject of the Unity of Religious Ideals, Feminist Philosophy, Non-violent communication methods and Social Psychology to many Universities in the globe. She has dedicated years of work assisting aid organizations including work with Mother Teresa in San Francisco and Calcutta and as director of the Prison Library Project. See http://www.sufimovement.net/rabia.htm for more information. Phone: (925) 254-2250. E-mail: aperez-chisti@itp.edu. Institute of Transpersonal Psychology, 1069 East Meadow Circle Palo Alto, CA 94303

way to enter a self-reflective reality. The use of transcendental phenomenology presented the criteria that provided correlation to my own experiences as part of a transformative feminist approach opening new ideas about intuitional similarities among women.

Hermeneutical approaches are valued as they offer concepts of bridging distance found in textual style. Hermeneutics as a research method possess a mysterious intimacy that grips our entire being, as if there were no distances at all, and every encounter with it embraces our entire being, as if there were an encounter with ourselves (Gadamer, 1976).

These three Holy Women were chosen for multiple reasons. They represent three distinctive spiritual paths: Mother Teresa of Calcutta represents the Christian religion, Nur-un-nisa Inayat Khan represents the Sufi Path, and the Amanuensis Reverend Frida Waterhouse was born Jewish but remained uncommitted to any faith system throughout her life as she believed all faith systems brought value toward spiritual emancipation. What unites these women culturally is the sociological marginalization of women during their historic timeframe. I do not examine the historic relevance of female gender repression; I simply mention what each woman found as

she experienced her circumstances. The emphasis of this examination is focused on their intuitive insights and my relationship to those experiences as I read their stories. I am interested in the synchronicity of arising intuitional experiences.

The fact that I knew two of these women, Mother Teresa and Reverend Waterhouse, became a key asset toward mapping correlation with my lived experiences. I learned about Nur-un-nisa through her brothers, Vilayat Inayat Khan and Hidayat Inayat Khan, with whom I was privileged to have long term relationships and who offered specific details about Nur's history.

I have come to understand a feminist approach as a pathway into unity in that we can only see what we ourselves can see at any given moment; even when we think we are reading about another's life, we are essentially reading about our own. My hope is that you read these stories as a gateway into yourselves, your insight, and your purpose. As colearners, I propose we begin approaching narrative material in a postmodern way by experimenting with multiple approaches that open vistas into hidden potential in ourselves while at the same time regarding our bodies' felt experiences as instruments of knowledge. This is the manner in which I explored transcendental phenomenology and herme-neutics.

As you approach these stories, you will read some of the salient pieces of information that give insight into the Holy Women's lives, correlated by the use of "consilience" (leaping together) of my own arising remembrance of an incident in my life. This is followed by defined criteria identifying a self-evaluative stance while reading the life stories. The inner/outer phenomena that guided synchronicity followed by my conclusion. My suggestion for beginning is to take some quiet time before you read these stories and invite your body to read with you.

The basic information on Mother Teresa's early life and development can be found in many texts and on many websites. Facts are synthesized below for a brief overview but the extraction for analysis was taken from a moment in Mother Teresa's early life cited in Katherine Sprink's (1997) biographical account.

Mother Teresa was born Agnes Gonxha Bojaxhiu on August 26, 1910, and she died on September 5, 1997. Agnes was born in Skopje, which today is a capital of the Republic of Macedonia. She was the youngest of the children of an Albanian family from Shkoder, born to Nikola and Drane Bojaxhiu. Her father was involved in the politics of the day and was devoted to the Alba-nian Cause. After a political meeting he fell ill and died. Agnes at the time was about eigth years old. Her father's death was shocking for the whole family as it caused fear about how the family was going to survive. Women during

that period were confined to a social life of wife and mother, and marriages were usually arranged within closed sys-tems of family negotiation. Her mother, seeking protection in her faith, directed Agnes to the Catholic Church. After her father's death she was raised as a Roman Catholic by her mother.

According to a biography by Joan Graff Clucas (1988), during her early years, Agnes was fascinated by stories of the lives of missionaries and their service. By the time she was 12, she was convinced that she should commit herself to religious life (Clucas, 1988, p. 24). She left her home at age 18 to join the Sisters of the Loreto as a mis-sionary. At that time the Loreto Sis-ters were a branch of the Institute of the Blessed Virgin Mary founded by an Englishwoman, Mary Ward, in 1609. Mary Ward had a vision for a different mode of religious life for women. She envisioned women living a life in com-panionship and discernment, inspired by the gospel and engaging with the world without the constraints of the traditional cloister, or an established 'Rule,' placing them under the governance of men. She also believed that women were equal to men in intellect and should be educated accordingly. These principles appealed to Agnes and set out to join the Loreto Sisters but she did not realize she would never again set eyes on her mother or sister (Sharn, 1997).

Agnes initially went to the Loreto Abbey in Rathfarnham, Ireland in order to learn English, which was the language of the Sis-ters of Loreto. They used English when instructing school children in India. Arriving in India in 1929, she began her novitiate in Darjeeling, near the Hima-layan Mountains. She took her first vows as a nun on May 24, 1931. At that time she chose the name Teresa after the patron saint of missionaries. She took her solemn vows on May 14, 1937, while serving as a teacher at the Loreto con-vent school in eastern Calcutta. Although Teresa enjoyed teaching at the school she was increasingly disturbed by the

poverty surrounding Calcutta. A famine in 1943 brought misery and death to the city, and the outbreak of Hindu/Muslim violence in August 1946 plunged the city into despair and horror.

Three years later, Teresa experienced what she later described as "the call with-in the call" while traveling to the Loreto convent in Darjeeling for her annual retreat. "I was to leave the convent and help the poor while living among them. It was an order. To fail would have been to break the faith" (Clucas, 1988, p. 35). It is here that we begin to realize the impact of the intuition on Mother Teresa's future choices. Although she identified herself as a Loreto, she knew this moment would separate her from the Loreto Sisters. She began her missionary work with the poor in 1948, replacing her traditional Loreto habit with a simple white cotton chira decorated with a blue border and then venturing out into the slums (Clucas, 1988, p. 39). The rest of her work is legendary and fills hundreds of books.

An important element of her "intu-itive-sensing" arises in the experience she details below. After her death, a diary was found with a description of the difficulties she faced at the begin-ning of her time in Calcutta. She had no food, income, or place to stay. She had to resort to begging and loneliness. This experience was a key factor for her feeling sense and awareness of what the poor were feeling and dealing with on a daily basis.

Mother Teresa proved how every revelation can become one's religion. Every intuitive process that integrated her physical form of knowing with her purpose became her spiritual path.

In Sprink's (1997) book, *Mother Tere-sa: A Complete Authorized Biography*, she referred to a comment Mother Teresa made about the doubt that rose up in her awareness including the desire to return to the convent:

Our Lord wants me to be a free nun covered with the poverty of the cross. Today I learned a good lesson. The poverty of the poor must be so hard for them. While looking for a home I walked and walked till my arms and legs ached. I thought how much they must ache in body and soul, looking for a home, food and health. Then the comfort of the Loreto (her former Order) came to tempt me. "You only have to say the word and all is yours again," the Tempter kept saying. Of free choice, my God, and out of love for you, I desire to remain and do whatever be your Holy will in my regard. I did not let a single tear come. (p. 37)

Knowing the plight of the poor, Mother Teresa's inner dialogue with what she called the 'Tempter' illuminates a bridge that brought two areas of knowing together. An awareness of the temptation for an easier life, protected by the church, revealed an altruistic understanding that motivated her actions. The scale of human suffering she saw around her and knowing she could not tolerate it personally activated the commitment within to serve and led her into her life's work. Instead of being repelled in hysteria toward safety, and helplessness in a male dominated religious structure, she took action. Even though the odds of receiving help from the Roman Catholic Church headed in the 1960s by Pope John XXIII brought little hope for change in women's relationship with the Church during the second Vatican council the recognition of women began to take on a central theme of exploration. Women were beginning to be recognized as valuable and were finally seen as members who could offer something besides money to the Church. Mother Teresa had the centers in Calcutta well established by the time the Church started paying attention to her projects.

Mother Teresa proved how every revelation can become one's religion. Every intuitive process that integrated her physical form of knowing with her purpose became her spiritual path. As Mother Teresa moved closer to her intuitional sensing of what was right for her core nucleus of being, her actions moved closer to her conscious action, and with sympathetic resonance, her structure of reality was altered. She felt connected to humanity, and her work moved her closer to God's creation: the human being who suffers.

If Mother Teresa found sympathetic resonance with homeless people, who she was compelled to serve because she too felt homeless, how much greater her story prepares us for finding consilience in all degrees of starvation in our lives.

When vision and action leap together a new direction is born in the heart of the individual. Individuals who risk following intuitive ways of knowing know without a shadow of doubt how they might miss fulfilling their life's mission. In Mother Teresa's case, she somatically understood suffering as the same experience to those she served. When she finally realized how untenable the condition of famine and homelessness was, she proceeded to dedicate her life to its prevention.

Consilience:

My body responds to Mother Teresa's story by bringing forward the remembrance of a particular incident of separation and loneliness in my life. This experience left an indelible cellular memory pattern in my body. It was not for a long time that I lived in the condition of homelessness and hunger, but it was long enough to frame an intersubjective apprehension of myself, my condition, and my belief system.

I had left a long-term marriage and needed to separate from my husband to understand the complex issues that were arising in our relationship. The experience of being alone without a comfort zone surrounding my lifestyle provided a detachment for me from an incorrectly perceived protected life.

Awakening to another day of feeling certainty about how the day would unfold was removed. The urgency of an empty belly, of not knowing where I was going to live, brought up feelings of desperation, loneliness, and shame. Hysteria and feelings of isolation caused a further distancing from others. I understood the pain in the physical body that Mother Teresa spoke about when she walked through the streets of Calcutta without finding any food or shelter. The time-space differentials do not enter ones complex somatic feelings when remembering being hungry. Otherness disappears, and only one experience remains intact. I thought how interesting it would be if we, as a global community, could enter a solidarity of hunger with those who suffer. What would happen if we dared to experience a similar or proximate condition? Subtle variances would arise in a study of this nature, but the body would certainly create a participatory synchronicity of reality.

Hungry, exhausted, and depressed, I walked further in the night talking out loud to myself, and demeaning myself for obsessive self-cherishing, I noticed a sign on the side of a building asking for help as a night clerk in what looked like a rooming house in San Francisco. I heard an inner voice say, "What is nourishment anyway?!" I was stunned by this inner question, as it seemed a bigger issue than just simply thinking of food to eat. Attempting to find an answer to this question, I stood there frozen staring into the building trying to see beyond seeing, anything that would make sense. For a sudden moment, I was standing in an open, cosmic spaciousness.

I had no information about this rooming house, which looked shabby from all outer appearances, but I followed my grumbling belly needing nourishment and comfort and went into the front office. I talked to the young man who ran the place. He hired me immediately and requested I work the night shift for no salary, but I would be given room and board in exchange for work. This arrangement was acceptable because it was immediate.

Working the night shift for several weeks, I noticed the rooms were mostly occupied by men. I would see police officers in the lobby during the evening hours that I had my shift. I came to understand that I had been hired on in a halfway house for ex-convicts.

My relationships with these men became extremely important to my devel-

oping understanding about incarceration. I later took a job as the Director of the Prison Library Project for five years serving the men and their families who were 'doing time on the inside.' I supplied spiritual literature for their libraries, read to them, and counseled them. Their stories were filled with regret and anger and reflected their stubbornness to change as they shared problems which brought them to self-imposed starvation of one kind or another.

My experience at the halfway house and Mother Teresa's momentum to act under oppressive circumstances gave meaning to my developing empathy. In Calcutta I gained transcendental insight into my experiences with men on parole who offered an accurate representation of a marginalized community. During a temporary period of my youth, I, too, was disassociated from members of my family and community. I could not return to the comfort of my marriage. I was

a prisoner in the cell of my illusionary separation, starving for nourishment, and now I was united with a group of men who came from a similar jail. As Bo Lazoff, the founder of the Prison Library Project has often said, "We are all doing time."

If Mother Teresa found sympathetic resonance with homeless people, who she was compelled to serve because she too felt homeless, how much greater her story prepares us for finding consilience in all degrees of starvation in our lives. I found a mutual sanctuary as well with those who had a similar experience of deprivation and famine but not necessarily the kind that Mother Teresa exposed to the world. Those who Mother Teresa served shared the same sense of abandonment by the societal class in which they were born, only hers was the Catholic Church to which she could not return to find comfort. In her simple act of service following her intuition, which told her not to return to the convent, she changed the world view of the poor by remaining in the same condition and served them from that mutual condition.

Nur-un-nisa Inayat Khan was born in Moscow, Russia, on January 1, 1914 to the saintly Indian Sufi mystic Hazrat Inayat Khan and Ora Ray Baker, an American from Albuquerque, New Mexico. Baker was the niece of Mary Baker Eddy, who began the Church of Christian Science. Hazrat Inayat Khan was raised in India in a family of musicians and Sufi Mystics; he was surrounded by

Orders. Although learned in the traditional Sufi Principles, his Message of Love, Harmony, and Beauty brought a new impulse to the West as he offered teachings on Spiritual Liberty and the Unity of Religious Ideals breaking away from conservative, doctrinal systems. He broke free of codified religious behavior that proliferated insularity. As he prepared his Victorian students to receive the Sufi Message that was years before its time, he married Ora Ray and fathered four beautiful children who all received an element of his inner awakened being. Nur-un-nisa was his first daughter.

Nur-un-nisa was regarded as a sensitive and shy individual with dreamy aspirations. She studied child psychology at the Sorbonne and music at the Paris Conservatory, specializing in harp and piano with the famous teacher, Nadia Boulanger. When her father died in 1927, she took care of her brothers and sister and took over the mothering of her mother, who retreated to bed in despair at the death of her husband. It was a very difficult time for her and the whole family. Yet during that period she wrote a well-published book on the *Jataka Tales* of the Buddha for children. The book is a valuable testament and insight into the heart and soul of this individual as she reveals mystical understanding absorbed through her father's profound esoteric knowledge.

When World War II broke out and France became overrun with German troops in 1940, the family fled to Lon-

When I read about Nur un nisa's resistance to leaving her post as radio operator and not go against her esoteric knowledge even under torture, this idea opened a deep question about how one might value dying for a purpose.

refined artists in music, art, and poetry who were trained in the great esoteric traditions of the Eastern world.

His esoteric work brought synthesis to four major Sufi Tariqats: the Suhrawardhi, Naqshabandhi, Qadiri, and Chisti

don. Nur-un-nisa was a pacifist by inherited teachings of her father and was deeply conflicted about how to help in this untenable situation. Her brother (Vilayat) recorded a conversation in which she conveyed her idea about wishing to act in some way that would bring benefit and not to collude with injustice through passivity. He stated,

On the eve of the war, Nur and I conferred deeply and at length on the pros and cons of our participation in the war. The problem was the same question asked today by conscientious objectors. We had been formed at the school of our Father, an Eastern sage and teacher. Behind him lay the entire tradition of Eastern spirituality. The then budding Gandhi inspired non-violent campaign had proven its effectiveness as a means of confronting violence but was barely explored in the West. And was this not the message of Christ? Was there not a contradiction in killing in order to stop manslaughter? But suppose a Nazi should hold hostages at gunpoint and starve them to death; it would be complicity to their murder if, having the means to kill the Nazi and unable to otherwise prevent him from carrying out his deed, we abstain from doing it in the name of non-violence. As we had that conversation, could we have ever imagined that one day Nur would find herself in the plight of the people she wanted to save? (Vilayat Khan, n.d.)

After being assigned to a bomber training school in June 1941, she applied for a commission in an attempt to relieve herself of the boring work there. Later she was recruited to join the F (France) Section of the Special Operations Executive, and in early February 1943 she was posted to the Air Ministry, Directorate of Air Intelligence, seconded to First Aid Nursing Yeomanry (FANY), and sent to Wanborough Manor, near Guildford in Surrey. From there she was sent to various other SOE schools for training, including STS 5 Winterfold, STS 36 Boarmans, and STS 52 Thame park. During her training she adopted the name Nora Baker.

Her training was incomplete, and her superiors held mixed opinions on her suitability for secret warfare. Nevertheless, her fluency in French and her competency in wireless operation—coupled with a shortage of experienced agents— made her a desirable candidate for service in Nazi-occupied France. In June of 1943 she was given a cryptonym "Madeleine" W/T operator "Nurse" and under the cover identity of Jeanne-

Marie Regnier, Assistant Section Officer/Ensign Inayat Khan was flown to landing ground B/20A "Indigestion" in Northern France on a night landing double Lysander operation, code named Teacher/Nurse/Chaplain/Monk. She was met by Henri Dericourt.

She traveled to Paris, and together with two other women (Diane Rowden, code name Paulette/Chaplain, and Cecily Lefort, code named Alice/Teacher) Nur joined the Physician network led by Francis Suttill, code name Prosper. Over the next month and a half, all other Physician network radio operators were arrested by the Sicherheitsdienst (SD). In spite of the danger, Nur rejected an offer to return to Britain, and she continued transmitting as the last essential link between London and Paris.

Moving from place to place, she managed to escape capture while maintaining wireless communication with London. "She refused however to abandon what had become the principal and most dangerous post in France . . . and did . . . excellent work" (Central Chancery of the Orders of Knighthood, 1949).

On or about October 13, 1943, Nur-un-Nisa Inayat Khan was arrested and interrogated at the SD Headquarters at 84 Avenue Foch in Paris. She appeared so gentle and unworldly that the SOE trainers became fiercely afraid of her and treated her as an extremely dangerous prisoner. She would say, "You cannot take from me that which you know not" (Stevenson, 1976). Her interrogation lasted a full month. She attempted an escape twice but was captured immediately in the same vicinity. She was taken to Germany on November 27, 1943 for safe custody and imprisoned in Pforsheim in solitary confinement with no contact with the outside world.

She was chained to the cement floor and tortured continually by being brutally beaten with a stick and whipped by a sadistic SS guard Wilhelm Ruppert before being shot in the back of the head. Her last words were "Vive La Liberte" (H. I. Khan, personal communication, July 15, 2007). Nur-un-nisa Inayat Khan was posthumously awarded a British Mention in Dispatches and a French Croix de Guerre with Gold Star. She was the third WWII FANY member to be awarded the George Cross, Britain's

highest award for gallantry not on the battlefield.

Consilience

Nothing could equal the difficulties that a person endured during WWII. This was a war founded on hatred and bigotry, and yet there are correlations of experience and many kinds of wars as everyone we meet is fighting a great battle. Wars are fought on many levels, but Martin Luther King reminded us how the moral universal arc is very long as it bends toward justice. This comment gave me hope as I read Nur-un-nisa's story. She helped me understand cognitively something about the "moral universal arc" from an episode in my life when I faced obstruction of justice at the point of a gun.

When I read about Nur-un-nisa's resistance to leaving her post as radio operator and not go against her esoteric knowledge even under torture, this idea opened a deep question about how one might value dying for a purpose. We will never know the exact circumstances of her experiences but hearing her story from her brother's information helped me understand a difficult decision that paired with mine when under a life-threatening situation. The following moment rose up in my mind's eye once again, and I felt the scene enter, like a moving picture, into reality.

I was riding in a truck in the Somali terrain with a team of other medical workers, bringing food and medicine into refugee camps. Our supply trucks were abruptly held up by young terrorists, who, while pointing guns at the entire group, asked all the U.S. workers to disembark. We did not know what would unfold, but it did not look good. I experienced our collective destiny going blank for one quintessential moment which felt like an eternity. Sitting on the side of the road we watched these young Somali brigands dismantle our trucks of the food and medicines. As I thought of the 25,000 refugees who needed these supplies, my thoughts tried to make sense of an ethical paradox. I was witnessing the same community of people rob from themselves.

Just the simple image of Nur-un-nisa's image in my mind's eye activated a sense of possibility in my traumatized thoughts. "Is this the time to face my own death?" I asked myself. "If I act in some way antagonistic to these young men, will I get myself killed?" I felt I

had to quickly come to terms with dying today, now or very soon. I understood there was not much time. What difference does it make if I do something or nothing? I was so perplexed, and my adrenaline was pumping. Just thinking I could do "something" brought a lessening of fear about being killed, a feeling that everything, even if I died, would be OK. This feeling seemed to bring a hidden relief to my physical body. I felt a sense of fulfillment or completion in doing work that I truly believed was good, and it brought with it a sense of fearlessness.

During the grueling hours of sitting in the hot sun, a leader of the young band identified himself. After moving into conversation with him with careful attention to making my last words on earth the best words I could muster under the circumstances, I blurted out, "What do you think about Bob Marley's music?" He looked as stunned as I was by the question, and he began to laugh. "I like Marley," he replied, and we began to talk about music. The musical conversation merged into negotiation, and we convinced him to take only half of the food and medical load. This was fitting negotiation as the young brigands did not have the trucks they needed, and killing all of us would have created an international incident of such political proportion they would not have had time to profit from the robbery.

Could the brilliance of that moment so precisely recorded in my memory, and the beneficial outcome with a 14-year-old boy—who had a gun but no future, no food to eat, and no home to live in—work out? I can only guess that his alienation had reached reflective awareness by showing him no future at all except a life of despotic actions. If, during that incident, my own feelings declared my days on earth were to end, could the same momentary synchronicity have been the reason that gave Nur-un-nisa the courage to hold true to her spiritual understanding when under threat of her life ending?

She was a radio operator for the French Resistance positioned in a precarious place so that the families of the village would not be killed by the Nazis. Her desire to serve others was stronger than her desire for her own safety. Or was it just coming to terms with the fact that her own death for a cause she believed was worth any sacrifice? Was not one life a worthy exchange if it saved many?

Knowing Nur's decision to remain in full integrity with her spiritual understanding, and paired to a time when my life was in tenuous hands of young, uneducated, unsympathetic boys, produced a viable option for success. The past and the present "paired" in an inextricable variation on a theme, and that was the threat of life under a gun. The circumstances were different, but my innate body remembrance held true to my intuitive feelings to respond the way I did. This response changed the course of action in a time when my life was at risk. I did successfully convey the rest of the medicines and food into the medical camp and return home safely.

Frida Waterhouse was born of Jewish parents on October 12, 1907 at Gloversville, New York, and she died on November 18, 1987. The year 1907 was the prelude to an economic depression in the United States, and there was barely any food in the house for the family of four. Both parents became very ill and the children were sent away to be cared for by adults who traumatized them with beatings, rape, and locking them in closets. When the family came back together a few years later, they moved to Los Angeles.

Although her mother and father

remained in fragile health, they both joined the Young People's Socialist League and brought the young girls with them to the meetings. Frida remained active in this organization until the age of 19. The discussions that took place at the meetings imbued her with a desire to serve her fellow human beings. She was stimulated to question authority and political leaders, and this activated a rebellious sense of identity. She had a difficult time conforming to the status quo found in feminine socialization of her time. The limitation and subjugation of a woman's role, which was relegated to serving a male and producing children, completely disgusted her sense of identity. She felt limitation when applying for jobs. Positions of authority were held by men, and she became fiercely angry about being talked down to when her intellect was far superior to those with whom she was applying for jobs.

She took odd jobs to survive and decided to have common law relationships instead of marrying. Later she did marry, but the relationship was not successful although it lasted 13 years. She attempted to follow the expected patterns for women and stay in the status quo, but she could not stand it. After separation and divorce she considered her divorce the single most important factor that moved her toward her spiritual goals. Although she enjoyed sharing stimulating discussion with her mother and father as the emphasis was on politics, music, and literature, her inner self-discovery became an important reason for her to separate from family and move to San Francisco to begin a new life.

Her anguish over the breakup of her marriage and the onset of a genetic inheritance of cataracts in both eyes caused blindness and emotional upheaval. She did not submit to surgery at that time because of the unknown factors surrounding the surgical procedure, but most importantly she was guided by her intuition to remain blind so that she could develop inner sensitivities. She attained the levels of spiritual insight born from her decision to remain blind for the duration of 4 to 5 years, a period she indicated developed her spiritual work. This period became clearer as her insight became keenly focused within (Waterhouse, 1974).

The nature of her work was to provide a practical springboard to help others adjust to active and reactive spiritual changes. She offered counseling to people between the years 1972 and 1986. The spiritual force fields surrounding an individual's Dharma (Spiritual-Soul) path were her focus when working in a one-to-one setting. She called her vibrant inner work a meeting of the Three Selves of our Being. The Selves included the un-grown-up child, the conscious self, and the High Self or Divine Mind. Her work was an integration of the ancient Kahuna teachings, published later by Max Freedom Long, and Christ teachings advocated by the Universal Church of the Master, which became the umbrella organization to allow her to function as a ministerial counselor.

The Universal Church of the Master was an organization founded in Santa Clara, California in 1908 that supported the study and diffusion of spiritual truths found in history, mundane and arcane sciences, spiritual and natural philosophy, occultism, astrology, psychometrics, parapsychology, and theology. The community respected pluralistic study but focused on the study of Christian principles as found in the Aquarian Gospel of Jesus Christ.

An incident that Frida conveyed to me one day during our 15-year relationship gives evidence to the power of her insight and intuition.

Frida suffered from scoliosis, a bone degeneration that curved the spinal column and caused her to remain bent and with a slight hump on the left rear portion of her back. After eye surgery that enabled her to see, she had to wear very thick glasses, which caused her beautiful clear blue eyes to appear quite large when gazing frontally at her face. She was very small (4'8") and thin, had white hair, walked with a cane, and could not see very well.

One day while waiting for a street light to change, a young man came up behind Frida and offered his arm to assist her in crossing the busy downtown San Francisco intersection. As they walked arm and arm she immediately felt her intuitive voice alert her to danger. She turned her face fully to him and, gazing deeply into his eyes, asked, "How can I help you?" He looked stunned and backed up, as though being stripped of all his clothing, and began to weep. They did not exchange any further words; she rather just gazed upon him for a long moment with great love and offered him one of the many pens she carried in her oversized purse to take down her phone number for a session. He scribbled it on his hand and quickly ran away.

After a few weeks he called her, and she invited him to see her at her apartment in the Mission. They began to talk, and he confessed his intentions that day on the street. He said he was going to rob her and steal her purse, but her response took him by such surprise, he could not do it. He confessed he cried for weeks. He said that her gaze caused him to see something greater in himself that he had never before seen. He knew he was a good man. Something in her eyes

These Holy Women's stories can inspire in us new epistemologies for learning how to read ourselves as women who strive for social autonomy and spiritual insight based on sacred relationships with ourselves and others.

validated his true nature. She confirmed this in their following sessions, which she offered to him without any financial exchange. He began to work for her doing odd jobs and secretarial help and became her most devoted student. His success in his professional life was confirmed in later years. Frida truly believed as the old Jewish proverb prescribes, "If you help one person, you help the whole world."

Hundreds of people came through Frida's door every year either by acquaintance or accident such as the young man who wanted to rob her. She met each person, one at a time in counsel. She formed women's and men's groups to study and work on collective issues arising from gender profiling. It is said in the ancient writings of the Sufi mystics that a wise person can arouse people to confess the secrets of their hearts, hidden thoughts, and feelings, which have made them ill in some way.

Consilience

I met Frida right after I experienced a great loss in my life. My son was born with malformed lungs, and at the time of his birth, there were no surgical solutions. I witnessed a prolonged death of an infant son who became more and more fragile over a seven-month period. As I watched him disappear from the human form, I harkened back to the time when he was in utero. I knew from the onset of the first trimester that something was wrong. Everything about my pregnancy held a signal about his formation going in a negative direction. It was as though I was witnessing a miracle of conversation within my body. I could not rationalize why I felt so oddly alert to something terrible about to happen. I cried continually trying to bear this burden. My impulse was to have an abortion. It was not a time when abortion was easily accessible. There were no clinics for information, it had to be done undercover or in a back alley arrangement. I fully identified with the thousands of women who suffer under this oppression.

There was so much chatter in my brain about ethical rights to life and religious intolerance that looked upon the act of abortion as criminal. Of course, I would repeat to myself continually, I was bad, I was a single woman who had sex outside of marriage, got pregnant, and now wanted to kill my baby. The relentlessness of my thoughts entered into all areas of self-diminishment and guilt feelings. Just having these thoughts made my emotional condition worse.

I knew I was going to lose him and at

the same time, I was more sensate and more present than ever before. Everything that my perceptive senses could pick up was amplified. I felt wildly disconnected from everyone and out of control and yet involved in an inner dialogue, a copresence so inherently spiritual everything joined in one sense of meaning and awareness.

Years later when I met Frida, I was still living in this complex state. I wanted to talk to a wise elder about my intuitive prebirth feelings. I had known intuitively that I needed to act on what I felt, but I did not follow through. I now sympathize with women who suffer when they do not follow their bodies' messages. I knew when I met Frida, she would bring clarity to my confusion.

She eventually found out all the details in our sessions together. Rather than giving me sympathy or what I would describe as outward messages of comfort, she began to tell me the story of her rape as a child instead. Her story detailed pain, fear, and self-diminishment—all the feelings I had. It seemed she was talking about all women's suffering and their difficulties in societies that do not take action against systemic forms of brutality accepted so casually in cultures around the globe. The absence of her parents' protection, the poverty during the depression era, and many other conditions found in her narrative story compiled the data for her sharing.

She insisted I listen carefully to her story. Her sharing eventually derailed my sorrowing because as she began to speak, I began to access a bridge to myself. I was beginning to sense similar qualities as in "mutual participation" with her life experiences joining with my own intuition and "felt sense." I could understand through her suffering, limitation, and difficulty how these feelings represented a genuine framework of acknowledged and sustained oppression in the lives of women. How she unraveled these tangled knots of awareness was a helpful companion for gaining connections within myself.

Mandala by Sharon Till

But the manner of her intuitive insight, which illuminated Frida's story for me, was the way she listened to her inner visions that led her to seeing when she was blind. It was during this period that she grasped the inner details of her childhood rape. Her sense of loss of control, the pain and the pleasure, all confused in a child's mind, alerted me as it correlated to deep feelings in my emotional body about being out of control yet in total intuitive communication with my child. All those feelings that I had experienced within myself were given 'copresence' with another living being. I felt not only confirmed about my ability to read what was going on in my body, I felt united with all women. It was an amazing universal knowing, a validation of sorts, that acknowledged a feminist epistemology, the woman's body as a source of wisdom. I just had to learn how not to shut it off and fall into self-pity.

Frida's overview was detailed in the way she understood "control" and how this concept is often misunderstood. She related control to surrender to the Divine Will. She indicated this did not mean surrendering to a person, per se, but rather how, through any relationship, we can learn vulnerability and therefore develop an awareness of a greater surrender with what she called "a finite being." If we have no control of the growth of our hair and nails, our birth and our death, why do we assume that we can control the things that happen to us? Any learning experience that arises in our lifetime is a gradual process from overt to more subtle nuances. It takes great courage to totally release our human will to Divine Will because it is a radical way of living. She conveyed how our human desires grow less important as we begin to understand the value of listening and acting from our inner intuitional voice.

The pain that I felt held a paradox only because I did not follow what I knew to be a right course of action from my understanding. An action that truly follows intuition is sometimes a terrible test of faith. It is not predicated on what society decides is right for a woman. Only I knew what needed to happen, and I needed to believe I could take the responsibility of such an action in the best sense of ethical benefit for myself and my child. I recognized the sociological moral stalemate that sets women up not to follow what they know from within their own bodies but rather to follow what is outside their own body as the rule that determines their action.

The Approaches Using Aspects From Different Methods

Transcendental phenomenology (Moustakas, 1994) offers a viable structure for connecting the inner and outer aspects of a narrative to any researcher's experiences. The feminist approaches, although widely varied, advocate proximity of inner motivation to social action (Lugones & Selman, 1990, p. 21). Feminist approaches takes into account women's intuitive feelings, their personal view of the world and themselves, their private concepts, their strivings for self-actualization, and their contribution to social justice; I can use feminist methods to correlate them with my own life. Transcendental phenomenology coheres well with hermeneutical reflection because it fulfills the function that is accomplished in bridging information to a conscious awareness (Gadamer, 1976).

The intriguing correlation of women who lived in the past and the uniting of the same phenomena that emerges in my own consciousness define the inherent alchemy found in narratives of great souls that identify the humblest aspects of human intuitive experience as an intimate and mysterious meeting ground. Edmond Husserl (1977) defined the term copresence as "the Other and I would be the same" (p. 109).

Similar aspects of unity are found in cognitive psychological theories as developed by Jean Piaget (1932), and Sigmund Freud (1959), who called attention to active ways in which the mind generates similar meaning and experiences. We know from phenomenological observation, as Ulrich Neisser (1967) stated, "Whether beautiful or ugly or just conveniently at hand, the world of experience is produced by the [woman] who experiences it" (p. 3). Neisser suggested a potential and unique insight into the "complexified" (Gebser, 1949, 1953, 1985) world of meaning that can arise from a unique experience.

I gave attention to the values and uniqueness of each of these three Holy Women's experiences, and I attempted to absorb the data through my feeling-sense as I paid attention to the sympathetic observation that brings my own arising awareness about myself into presence. My story functions as an indication of mutual participation through arising sympathetic awareness.

As I approached this process, I wanted to understand how these women gained new understanding from their struggles and how intuition played out in their lives. The narratives uniquely reveal a quadrant of intersubjective ways of bringing evidence forward about their intuitive processes found through the four qualitative criteria defined below: transcendental, pairing, copresence, and consilience.

Transcendental

Although the term "transcendental" might be assumed to mean going beyond one's experience, M. Farber (1943), and Clark Moustakas (1994) listed five descriptions that reveal a way to unite the terms transcendental and phenomenology:

1) It begins with the things themselves,

2) It is not concerned with matters of fact but seeks to determine meanings,

3) It deals both with essences and with possible essences,

4) It offers direct insight into the essence of things, growing out of the self-givenness of objects and reflective descriptions, and

5) It seeks to obtain knowledge through a state of pure subjectivity, while retaining the values of thinking and reflecting. (p. 568)

A natural question arises concerning the validity of what we see, feel, and sense and which might indeed be misunderstood as a fiction. One person might experience one thing while reading a narrative, and another person might find something else, but as the stories unfold and my story becomes intertwined with the narratives, what might be apprehended by yet another reader is a transcendental "fresh" (Moustakas, 1994) way of knowing many complex systems of interpretation. One or more of these millions of sensory stimuli could illuminate a sense of connection with spirit revealing yet another deeper meaning in how these Holy Women and the researcher might interface in a real-time, living synchronicity, the benefit being a generous and exponential gift of awareness for the researcher and the reader.

Pairing

Psychologists have spent years focusing on understanding how an individual perceives, synthesizes, and interprets feelings, world view, inner life, and the nature of subjective sensing (James, 2001). They describe multiplicities of experience that align with the idea of "pairing" (Moustakas, 1994; Kirby, 1989). Pairing is seeing the world that is within me from another's eyes. This term encourages us to look at the interrelated components that primarily form this communion of spirit and evaluate it as to whether another's experience informs my own, cognitively and experientially.

There are obvious limitations that should be considered at this point. Women who are historical figures, such as Mother Teresa, Nur-un-nisa Inayat Khan, and Reverend Frida Waterhouse, are no longer living. The intersubjective communication between persons who test out their knowledge of each other in order to understand each other is one sided. Another limitation found in narrative data collected from secondary literary sources such as biographical or ethnographic material can include inferences from others and not necessarily reflect the historic person. As the women are not living to validate elements of their experience, a feminist method offers a suggestion for the integration of these variables.

In qualitative research, particularly empirical phenomenology, distance from the subject is recommended when the researcher examines the same description repeatedly and delineates each time a transition in meaning is perceived with respect to the intention of discovering the meaning (Giorgi, 1979). The researcher then eliminates redundancies and develops meaning units.

Through feminist approaches, similar meaning units might occur, but they would be assembled through the attempt to integrate the distance between the researcher and the variety of sources about the one being researched by regarding the proximity and correlation of phenomena instead. This is done through the aspect of copresence.

Copresence

The guidelines for observation are found when the researcher acknowledges the biases that can exist, through what may be termed in feminist theory, as an intuitive experience or a copresense with another body of information coexisting with my body of intuitive experience. (Braud & Anderson, 1998). A communion arises in the researcher who is directly present to a description as defined in the Latin "communionem," meaning "mutual participation." The idea introduces the researcher to the task of identifying the phenomenon of an intuitional experience and correlating it to the same intuitional experience living within the researcher. This will bring about the aspect of consilience.

Consilience

Synchronistic correlation can be understood as a concept of consilience. This term was coined by the British polymath and (what we would call today) philosopher of science, William Whewell (1794-1866), in his 1840 publication, *Philosophy of the Inductive Sciences*. Whewell's consilience of induction occurred when an induction obtained from one class of facts coincided with an induction obtained from a different class of facts. This consilience was one of several tests of validity of the theory in which it occurred. Consilience means, literally, "to leap with" or "to jump together." In an explanatory surprise, two or more sets of inductions jump together and are seen to be importantly interrelated. (W. Braud, personal communication, August, 2008). Using Whewell's idea, the researcher can obtain a framework that explicates the research from various descriptions while addressing mutual situations.

The concept of consilience elucidates the feminist approaches to transcendental phenomenology and hermeneutic studies. By its inherent meaning, consilience can link these great women's contributions to religious thought and philosophy by bringing new relevance to stereotypic concepts that inform women's social behavior, particularly during times of historic isolation, oppression and physical incapacity such as during World War II, the Great Depression, and so on. The same correlation can bring fresh meaning to how religious codes of behavior and spiritual understanding of these codes can inspire individuals to take altruistic actions.

These Holy Women's stories can inspire in us new epistemologies for learning how to read ourselves as women who strive for social autonomy and spiritual insight based on sacred relationships with ourselves and others. The use of consilience might be a useful approach for a researcher who is are looking for substantive connections.

Merleau Ponty (1962) pointed out how the real is not to be constructed or explained but described. This is the appeal to one who uses narrative material to prepare research data that can bring to life universal expressions. Social psychologists have recognized the need to see greater influence of interpersonal relations and group membership through investigation of attitudes on group dynamics. The emergence of the roles of personality on group dynamics has given rise to information that shows how particular prophetic roles become important as they not only serve to integrate the individual but also affect the primary group and the collective behavior of the greater community. The best examples of this can be found in Mother Teresa, Nur-un-nisa Inayat Khan, and Reverend Frida Waterhouse's life stories.

However, when a researcher experiences a living person a similar understanding abides, as Amedeo Giorgi (1985) cautioned us to be aware, "that a human subject is also historical and social and dwells in a world of meanings" (pp. 442-444). Whether we hermeneutically dig down into a narrative or collect data from a living person, the idea of Whewell's consilience was edified by Abraham Maslow (1965) who said, "Every person is, in part, [her] own project, and makes [herself]" (p. 308). His assertion indicates how an individual is fastened to a group and the group to an individual. As a researcher explores her arising experiences while hermeneutically culling data from textual sources, the researcher matures by developing a unitive sense of self and sees the world more as a reflection of one's own nature. Correlation produces a valid awareness of one's self as the enlargement of self is dependent upon and in turn supports the breadth of community values.

Ponty's, Giorgi's and Maslow's acknowledgment of experience can reveal hidden categories and codes that determine an inner exegesis—a unitive awareness that rejects subject/object valuing alone. Combined and respecting feminist approaches, the researcher can ride out the flow of intuitive "leaping together" of experiences and arrive at a new realm that gives access to deeper meaning about oneself and others. This idea can also be achieved when phenomena of historic individuals' decades past are found operating in the researcher's present-time psychospiritual momentum. Feminist praxis is committed to exploration in these areas. The phenomena that I considered consisted of three inner phenomena that expressed the internal physiological sensations and questions that arose in myself, while reading the Holy Women's stories. The three outer phenomena reflect issues that united women's experiences and who represented all women's condition in the world.

The following are the Inner Phenomena.

1. Did I maintain faith and confidence to follow my arising intuition?

2. Was my inner core in integrity with my course of action?

3. Was I able to understand the value of spiritual unity under the arising circumstances of social injustice?

The following are the Outer Phenomena.

1. Were they able to make the defining choice to live from their intuitive knowing even when moving against religious laws, oppressive domination through violence, or threat of bodily safety?

2) Were their levels of sociological marginalization, prototyping, or suffering evident in their historic timeframe?

3) What were the oppressions/threats that motivated them toward beneficent acts?

A feminist epistemology such as this avoids solipsistic circularity. As Shultz (1967, p.106) clarified, if I look at my whole stock of your lived experiences and ask about the structure of this knowledge, one thing becomes clear: This is that everything I know about your conscious life is really based on my knowledge of my own lived experiences. My lived experiences of you are constituted in simultaneity or quasi-simultaneity with your lived experiences, to which they are intentionally related. It is only because of this that, when I look backward, I am able to synchronize my past experiences of you with your past experiences.

The four approaches for self-examination as research deepens correlation with six phenomena linking to my experiences. They emerge up through a reduced attitude, meaning I sought out the essence of each phenomenon through the application of free imaginative variations while respecting the fact that my conscious choices reflect the sum total of themselves. As William James (1892, 1961, 2001) said, it is not only [her] body and psychic powers but "in the widest possible sense, a [person's] Me is the sum total of all that [she] can call [hers]" (p.58). I remained in connection with these Holy Women's experiences as they searched for themselves as human beings. How I sustained connection with these Holy Women's narratives was based on the following four approaches of self examination.

The Four Approaches of Self-Examination for the Reader

The Description — The Researcher's Biases

As long as I acknowledge my biases as I reiterate the description of the narrative and confirm the validity of my experiences with the narrative experiences of this study, the feminist perspective is respected and solipsism is reduced. The focus of what women face universally as oppressive centralizes the study. My unique awareness of my historic timeframe and oppression of women during my lifetime requires that I remain astute to the meaning and the limitation stated in the Holy Women's stories and what these differences bring to my felt-sense. The "leaping together" of accumulated data, and suggested feminist intuitive

approaches to interpreting experiences gather evidence of a valid, collective, "social experience of women" (Maquire, 1987, p. 39). The correlation of my experiences with the Holy Women's narratives may bring forward inherent biases that can arise in interpretation. Biases are acceptable as long as they are stated.

Reduction — The Researcher's Self-Observation Leading Back to the Source of Meaning

Self-observation begins with openness to the boundaries between the person being studied and the psychological condition of the researcher as a united observation. There is no prejudgment of what is being stated and observed. Open observation between the time/space abstractions among generations could reveal an existing experiential link as the physical/psychic world remains in a permeable reality. This criterion addresses how a unitive reduction described in feminist transcendental phenomenology removes the observational dualism of EMIC/ETIC stances. This can happen because the transcendent process makes known the intersubjective perceptions of what is real and unites both internal and external observational realities by enhancing reduction. Reduction directs us back to the source, meaning and existence of the experienced world (Schmitt, 1968, p. 61).

Search for Essence — Maintenance of Respect and Acknowledgment of Cooperative Partnerships

Acknowledgment of the experience of a Holy woman who is dedicated to repairing social inequity requires a prudent person to deliberate well about what is good or expedient for herself, not with a view for some particular end such as wealth or power but rather for the purpose of living well and being as fully awakened as she can be. Therefore, the researcher's respect toward an individual's spiritually dedicated life forms the initial approach to the subject. Then one approaches the search for essence through the artful idea of bringing something descriptive into being, "whether these things can be or not be, the cause of the production lies in the producer, not in the thing itself which is produced" (Aristotle, trans. 2005, p. 129). The mul-

tiple patterns arising from the various narratives correlated to the researcher's experiences shape the phenomena and the ethical attitudes of the researcher. The search for essence and one's ethical motivation to remain consistently engaged with oneself and one's development builds the respect for cooperative partnerships within the study itself.

Intentionality — Remaining in a "Present Intuitional Connection to Others" as Values Leap Together

Remaining present to feelings through an "intuitive-sense" arising through consilience is a primary practice for feminist transcendental phenomenology as it opens a connection to others. Values of a researcher become integrated into the research process when experiences from a specific woman's life narrative "leap together" and form a sympathetic resonance that ultimately influences another way of negotiating the structure of reality for the researcher that has not been known before and can be discovered, as Husserl indicated, through the real event that can only be discovered by being lived. Bergson (1946) verified this idea by calling things that cannot be grasped through immobile concepts as intuition. "In principle, there is within me a realm of virtually infinite access to other human beings" (Moustakas, 1994, p. 37)

"The very core and nucleus of our self-knowing is the very sanctuary of our life; it is the sense of activity which certain inner states possess. This sense of activity is often held to be a direct revelation of the living substance of our Soul" (James, 1892, 1961, 2002, p. 48). The women's races, cultures, ages, religious devotion, abilities, and intuitive understanding provide insight for the researcher's unique spiritual awakening as well as offer practical ways of addressing reality in the moment.

Differences need not be seen as distractions but rather as forces awakening the creative mover behind the unique heart of being. As the researcher and researched individuals draw closer together, maintaining care for this precious wealth in an arising integral synthesis, the intuitive ways of knowing within the context of how experience captivates the moment will be the determining forces of the researcher's fate.

Conclusion - A Participatory Process for Developing Feminist Research and Intuition

I hope these Holy Women's stories and the arising correlation to my own experiences offers a view into newer and better approaches, methodologies and strategies that will liberate women's experiences. The details in Mother Teresa's life work which reveal her resolute commitment to follow her inner voice, Nur-un-nisa Inayat Khan's bravery and the strength against oppression, and the fortitude which Reverend Frida Waterhouse embraced when dealing with blindness show us not only pathways to our intuitive sensing but, more than that, they challenge us to be feminists in our research. The challenge comes through our own arising awareness which is grounded in the felt sense of our bodies as instruments of knowledge in our willingness to read about another person's life while reading our own. Our arising thoughts and insights shared among researchers can actively enact a change in research approaches grounded in somatic reality of women.

As a feminist researcher, insight into the imaginative variation of my own experiences that correlated to the same phenomena of other women's experiences brought to light the living relationship of our humanness. Taking a position with respect to any human situation or real event has consilience only when being lived in a transcendental, paired and copresent manner since the lived experience manifests a pre-predictive unity of our life in the world.

The intentionality of these Holy Women's experiences show a direct intuitive sensing toward benefitting others even though their experiences were very different. The ethical and unfolding essence of consciousness itself was leading to a concept of unification. Feminist approaches offer significant strategies for research and investigation. Transcendental phenomenology and hermeneutics offers a fresh and open approach to textual descriptions, meanings and essences (Moustakas 1994). This method of research can bring consciousness-raising to all those who are attracted to participatory understanding as the method invites you, the researcher, to be a part of the research. I found the ability to bridge across race, age, historical, sociological, and faith based issues. Feminist approaches does not promote dominant paradigm views of the research which idealize separation of research and researcher nor does it look for redundancies but rather involves an activism that unites the search for essence in all concerned.

Finally, feminist approaches to transcendental phenomenology are not meant for women only. A synthesized approach of this kind, bridging researcher and narrative, does not negate the great narratives of men who function from intuitive knowing and act accordingly. As feminist scholars indicate, "research will proceed from a perspective that values women's experiences, ideas and needs rather than assuming we should be more like men." (Weston, 1988, p. 148)

My interpretation was offered as a pathway into direct looking at the unifying forces of those who function from intuition and act in dedicated service to humanity.

References

Bergson, H. (1946). *The creative mind: An introduction to metaphysics*. New York: Kensington.

Braud, W., & Anderson, R. (1998). *Transpersonal research methods for the social sciences: Honoring human experience*. Thousand Oaks, CA: Sage.

Central Chancery of the Orders of Knighthood. (1949, April 5). *The London Gazette*, 38578[Suppl.], 1703. Retrieved from http://www.london-gazette.co.uk/issues/38578/supplements/1703

Clucas, J. G. (1988). *Mother Teresa*. New York: Chelsea House.

Farber, M. (1943). *The foundation of phenomenology*. Albany, NY: SUNY Press.

Freud, S. (2009). *On creativity & the unconscious: The psychology of art, literature, love & religion*. New York: Harper Perennial Modern Thought. (Original work published 1925)

Gadamer, H. G. (1976). *Philosophical hermeneutics*. Berkeley: University of California Press.

Gebser, J. (1985, 1991). *The ever-present origin*. Athens: Ohio University Press.

Giorgi, A. (Ed.). (1979, 1985). *Phenomenology & psychological research*. Pittsburgh, PA: Duquesne University Press.

Husserl, E. (1977). *Cartesian meditations: An introduction to metaphysics*. (D. Cairns, Trans.). The Hague, Netherlands: Marinus Nijhoff.

James, W. (1892, 1985, 2002). *Psychology (briefer course)*. Notre Dame, IN: University of Notre Dame Press.

Lugones, M. C., & Selman, E. V. (1990). Have we got a theory for you!: Feminist theory, cultural imperialism and demand for "the women's voice". In A. Y. al-Hibri & M. A. Simons (Eds.), *In hypathia reborn: Essays in feminist philosophy* (pp. 18-33). Bloomington: Indiana University Press.

Maquire, P. (1987). *Doing participatory research: A feminist approach*. Boston : University of Massachusetts.

Maslow, A. H. (1965). *The psychology of science*. New York: Harper & Row.

Merleau Ponty, M. (1962). *Phenomenology of perception* (C. Smith, Trans.). Boston: Routledge & Kegan Paul.

Moustakas, C. (1990). *Heuristic research: Designs, methodology, & applications*. Newbury Park, CA: Sage.

Moustakas, C. (1994). *Phenomenological research methods*. Thousand Oaks, CA: Sage.

Neisser, U. (1967). *Cognitive psychology*. New York: Appleton-Century-Crofts.

Schmitt, R. (1968). Husserl's transcendental phenomenology reduction. In J. J. Kockelmans (Ed.), *Phenomenology* (PAGE NUMBERS MISSING). Garden City, NY: Doubleday.

Sharn, L. (1997). Mother Teresa dies at 87. *USA Today*. Retrieved from http://www.usatoday.com/news/mothert/mother01.htm

Shultz, A. (1967). *The phenomenology of the social world*. (G. Walsh & F. Lehnert, Trans.). Evanston, IL: Northwestern University Press.

Sprink, K. (1997). *Mother Teresa: A complete authorized biography*. New York: Harper Collins.

Stevenson, W. (1976). *A man called intrepid: The secret war*. New York: Ballantine Books.

Khan, V. (n.d.). Memories of My Sister. Retrieved July 8th 2009 from: http://www.angelfire.com/co/begumnoor/vilayat.txt

Waterhouse, F. (1974). *Why Me?* San Francisco: Rainbow Bridge.

Weston, M. (1988). Can Academic Research Be Truly Feminist? In D. Currie (Ed.), *From the margins to the centre: Selected essays in women's studies research* (pp. 142-150). Saskatoon, Canada: The Women's Studies Research Unit, University of Saskatchewan.

Whewell, W. (1840). *Philosophy of the inductive sciences*. London: John W. Parker, West Strand.

Photo Credits

Mother Teresa: Frangsmyr, Tore & Adams, Irwin (Eds.) (1997). Nobel Lectures, Peace 1971-1980. World Scientific Publishing Company. Singapore. Retrieved July 6th 2009 from http://nobelprize.org/nobel_prizes/peace/laureates/1979/teresa-bio.html

Nur-un-nisa: Khan, Vilayat. Sufi Order International. Retrieved July 6th 2009 from http://www.sufiorder.org/noor.html

Reverend Frida Waterhouse: A.P. Chisti, Copyright 1986

Integralizing Goddess

A Philosophical & Practical Approach to Spiritual Awakening

Through the (R)evolutionary Five-fold Feminine Force

Chandra Alexandre

I n seeking the Divine today, many are looking beyond their own backyard and finding joy, healing and a deepened connection to a yearning within by virtue of spiritualities and religious traditions that acknowledge the fullness of God/dess. In part because of a greater global awareness created through the strides of feminism, activism and communication networks, the impact of the Divine Feminine/Female today on individuals working outside of as well as within normative structures and faiths is becoming not only more discernable, but also more powerful.

With this in mind, Goddess is becoming a "transaction between contexts, encouraging us to discover how the two contexts [of human and divine] can illuminate each other in new and sometimes unexpected ways" (Harman, 1989, p. 4). For many, She is a radical role model helping those seeking expanded cos-

mogonies, cosmologies and consciousness break through limiting fears often inspired by superstition and ignorance. She is helping us overcome the biases of the western sociocultural milieu.

One only need look to the goddesses of India for an example. So great has been Her influence throughout the ages that borders have never contained Her. From Parvatī to Kālī, Ṣaṣṭhī to Sarasvatī, new versions of the goddess, new reasons for her existence, and even newer forms have appeared are constantly appearing. She has historically crossed with trade routes and migrations into neighboring lands, becoming a powerful and revered deity in her newly chosen homeland, wherever that might be. In both Tibetan and Nepali Buddhist contexts, for example, the Indian Kālī and Chinnamastā early on became Tarā and Vajrayoginī, respectively. And today, Devī (Goddess) is beginning to find recognition across oceans in the minds and hearts of those located in more distant foreign lands, most notably North America, Australia, and Europe.

Here in the United States, She is inspiring a (r)evolution of worship and is now assuming revised, context-specific meanings and alternative significances worked through the vehicles of individual bodies, minds and hearts. From the inside out, coming to us in dreams, meditations and through the

channels of sacred spaces, the goddess' archetypal symbols and forms are lifting us through and back into ourselves. She is in fact heeding our heart-song, flying through time and space back into our consciousness, refusing to be ignored. It would seem that our very souls are asking Her to emerge and She is heeding the call.

My work in this arena is at the intersection of the goddess-centered, embodied spirituality of India known as Śākta Tantra and the earth-based, goddess-centered spiritualities of the developed west. Śākta Tantra itself deeply resonates with much western reclamation of the divine female and itself is a system whose theology reveres the immanence of spirit. It in turn provides ritualized as well as lived spaces and practices in which the body acts as, and becomes substance for, the Divine.[1]

Śākta Tantra considers Female/Feminine energy to be fundamental and activating. Called śakti, this energy is the source of all creation without which the gods and humanity would perish. The religious Śākta texts, known as the Tantras and sometimes called the fifth Veda, expound upon this worldview. These works are renowned for doing so by illustrating philosophical concepts with bold sexual imagery. For example, Tantrick images of Goddess Kālī often portray Her standing upon the prone body of Śiva, her consort. Sometimes he

Chandra Alexandre, Ph.D., D.Min. is the founder of SHARANYA (www.sharanya.org), a devī mandir (goddess temple) based in San Francisco with a worldwide presence through Kali Vidya (www.kalividya.org), an on-line mystery school. She is an hereditary Witch, Aghori Nath, and the Chair of the Maa Batakali Cultural Mission in Puri (Orissa), India, where she received her first diksha (initiation) into Kali's mysteries. Chandra holds a Ph.D. in Asian & Comparative studies and a Doctor of Ministry degree in Creation Spirituality as well as an MBA in sustainable management. Contact: 2063 42nd Avenue, San Francisco, CA 94116. Tel: 415-505-6840. Fax: 415-723-7299 is ithyphallic; sometimes She is engaged in intercourse with Him. For Tantricks, not only does this depiction graphically and unmistakably portray the activating principle of śakti, but it underscores the importance of the relationship between Sacred Feminine and Sacred Masculine.

Through the ancient traditions of Goddess worship still practiced today in Hindu South Asia, women and men without religious or cultural access to the Divine Female and Divine Feminine may achieve a glimpse at possibilities for the Feminine and female beyond patriarchal limitations. For within some of these traditions are to be found expanded notions of divinity, religiosity and spiritual leadership that may aid micro and macro transitions, those that facilitate the leaps of consciousness and broadenings of awareness required to create sustainable post-patriarchal outcomes.

For example, through the fierce extremes of the goddess' paradoxical nature (as garnered from juxtaposition of Her demon-slaying tendencies in the Devī-Māhātmya with Her functioning as supreme reality in some of the Tantras, or with Her position as universal, compassionate mother in the devotional bhakta traditions), we may break free of dominator hierarchies that value and serve only one half of Creation. Such knowledge can make Her an impetus for expanded conceptualizations of both personal divinity and cosmic reality for anyone wishing to move beyond culturally-infused stereotypes of women/Goddess or beyond monotheistic traditions that subjugate, disempower, or otherwise deny the roles and rights of women and the Feminine on transcendent as well as mundane levels.

Additionally, for those seeking to awaken the intensity of the Divine Feminine in a land that has so culturally

[1] Although the Śakta and Tantrick traditions are often conflated because of the importance of the goddess within Tantra generally, the followers of Tantra are not all primarily worshippers of the goddess. There are distinct paths within Tantra; however, it is Devī as this fundamental reality who engages the heart, mind and soul of Śākta devotees.

and spiritually denied it, the presence of a goddess who manifests as intimate mother and transcendent, universal creatrix; virgin and celestial lover; faithful, complacent wife and bloodthirsty, independent huntress, as does Devī, can be reassuring. In any of these forms or combinations, She encompasses and transcends Western notions of duality, thereby defying our ability to quantify, qualify or explain Her existence in terms we might easily understand by virtue of our dominant worldview. Devī, it would appear, is beyond succinct and simple explanation. She is beyond a monological interpretation.

Today, a robust and dynamic complexity is being achieved through new traditions in which the goddess' symbols and rituals—some newly constructed— inform by connecting to passions, devotion and a desire for engaged spirituality that can not be contained by a deity's or a devotee's country of origin. Today, Goddesses of both east and west,

Today, the Goddesses of both east and west are coming together in order to reclaim the power of womanhood and the Feminine from within the facts of nature, our bodies, and the universe.

ancient and modern, are coming together in order to reclaim the power of womanhood and the Feminine from within the facts of nature, our bodies, and the universe. Through the Female Divine and Her agents empowered, subordinated groups and individuals are finding ways to reclaim marginalized ideological and literal spaces, including those of women's bodies, through new spiritual traditions and expressions that allow Her to reverberate throughout more than the immediate context and situatedness of subjects and communities. Even if she has not yet appeared in your purview, may this article whet your appetite for engagement with the delicious and difficult work She inspires.

Who is She?

The Dark Goddess embraces all. She

is a real and meaningful participant in the unfolding story of human evolution, a story upon which a number of authors in the west have already commented (for example Begg, 1985; Woodman, 1985; Baring & Cashord, 1991; Galland, 1991; Matthews, 1992; Gustafson, 1990; Woodman & Dickson, 1996). She is the totality of beingness and that which lies beyond. Her power—and why so many spiritual practitioners may avoid her— lies in our fear of the unknown and the antinomian. And this is not only our individual, personal fear, but also our collective, societal, and cultural fear. The magnitude of her power lies in our projections, with roots in our psycho-spiritual baggage. It lies where we are afraid to look because looking hurts or disgusts us. It lies within us because we are afraid to engage with the feelings brought on by secret desires, selfishness, disowned material, or what we believe (for one reason or another) is wrong, impure, or bad. When we look at the Dark Goddess, we see what we wish to keep out of awareness, and She terrifies us—for She is all that, including and transcending as She moves us, works us, catalyzes us not only into tomorrow, but into and out of ourselves. She does not hold back Truth, and for this, we are often terrified of Her. What might she reveal?

The power of the Dark Goddess lies in her claim to what we ignore. Darkness, indeed, is a metaphor for the deeper truth of our uncomfortableness with the full spectrum of energies the Divine offers. No, life is not mere light and fluffiness. We all know the world is filled with beauty and pain, ecstasy and horror. Look at Kālī, for example, Hindu Black Goddess, Devourer of Time. With her right two hands she presents mudras (hand gestures) signifying the granting of boons and blessings, and with her left two hands she holds a severed head and a machete. To look upon her, we in the developed world might wonder how to achieve the benefits and avoid the blood and gore. But Her revelation lies in and through the acceptance of her totality—

for the blade and head are in truth the goddess offering the promise of detachment and release from the constraints of our egos and limitedness. To take only half of Her is to miss the whole point; and She not only potentiates, but is, an integral wholeness.

Still, many of us (and many of our institutions), are just not ready to deal with what the Dark Goddesses have to say about the world, our place in it, and the nature of the sacred—let alone the spiritual journey. The realities the Dark Goddess so directly and honestly reveals can be harsh, challenging, and overall, can point to the necessity of making changes on the road to upper-level or integral memes and beyond—and these changes are usually difficult ones, the ones we'd rather think were not necessary. She lurks, just there, waiting for us to have the courage to reveal everything. Yes, she asks for our greatest vulnerability, and She demands personal responsibility. No wonder we are more than afraid.

In the 21st century, Her images have in some ways moved beyond the original projection in various cultural contexts of male fears (of the power of women, of birth, of sex, of death, and of the realms of the unknown and that which is judged through the patriarchal lens to be 'nasty,' which usually translates as that which the female and the feminine inhabit and represent), and instead taken on the mantle of the oppressed. She has arrived in our consciousness to empower all who suffer tyrannies and injustice. We can look to the mestiza (meaning mixed, as in of both Spanish and indigenous origins) goddess, Our Lady of Guadalupe, Patroness of America, to find such an example. Guadalupe, a mestiza and carrier of old and new wisdom, is a healer, miracle worker, and a symbol of the Mexican struggle for independence and freedom.

Even when Her image is not black or brown, however, we still may know Her as Dark Goddess, easily identifiable through myth, dream, intuition, and iconography because She oozes a terrifying fierceness—but one filled with compassion that can set our minds reeling at the seeming paradox. No, she is not just a demon-slaying, sword-carrying, freedom-fighter of a goddess. She is also sweet and tender, compassionate and graceful, comforting and facilitating of the greatest liberation through ahimsa (non-violence).

She is a role model for an appropriate response to injustice. In Her, we can not

Photo: Chandra Alexandre

only seek refuge through the transformational powers she possesses, but also find sanctuary in her wisdom, especially in moments of despair, hopelessness, overwhelm, sorrow, and loss. We can find reassurance during times of crisis because She holds this part of reality too—and holds it with loving kindness.

Working with the Dark Goddess

She can help us find the resolve, dedication, strength and faith to enter the abodes of fear. How can we come to know this Reality?

1. Intention

The first step on the path of engagement with the Dark Goddess is the setting of an intention. Called sankalpa in yogic traditions, this is our resolve, our commitment, our statement of yearning in the dance of free will and determinism. We must be clear in our call, direct and unpretentious in our approach. This is not a mere matter of reading words, however; this is the reality of taking steps toward the deep end of the pool. Self-reflection is key: Is your goal to reach the other end of the pool or just the rope in the middle? Are you willing to both swim gracefully and doggie paddle awkwardly as you reach for it? Do you know if you're sinking to call for help? What if someone leaves the heater off or turns the lights out accidentally? Will you go on? The more clearly you can articulate for yourself the nature of your exploration and where you wish to go, the more carefully and precisely you will be guided toward your destination. With a sankalpa is set, we are ready to begin.

2. The Five Qualities of Dark Goddess

Every Dark Goddess shares five essential qualities. They are: embodiment, relationality, cyclicity, the chthonic, and the antinomian. Through them, she offers us Her gifts, with liberation the greatest of them all. These five qualities are the very core of what it means to be a dark goddess, and thus, in concert they create a vibrational energy evocative of Her essence. In this way, your opening to the Dark Goddess is facilitated by working with them—they set the stage for a personal invocation that allows the Dark Goddess to manifest as one desires to know Her.

This may be the time to create a Dark Goddess altar or ritual space using the tools, symbols and practices offered here, inclusive of personal inspirations. Certainly, this can be purely an internal exercise if one desires. Regardless, it is recommended that one study these qualities along with their correspondences, some of which are garnered from Tantrick tradition, in order to feel both physical and subtle levels of manifestation; for these can help not only deepen spiritual discovery and self-awareness, but also help us to come closer to the heart of Her—and through this, closer to the heart of the world.

i. Embodiment

This quality is the root of all others. Literally, it corresponds to the Muladhara Chakra, the energy center at the base of our spine, the root of the body. Without this quality, there is little space from which we can engage Her.

To start, I will add mine to the chorus of voices articulating the many problems arising from a dualistic (Cartesian) worldview that deems the rational and intellectual more valuable than the body's wisdom. This wisdom, inclusive of the internal environment of emotions, intuitions, and sensations, was relegated to the abode of the unimportant and not-to-be-taken-seriously quite some time ago. While there are always benefits to be obtained from a concentrated focus through a particular lens, this worldview has created a variety of ways of looking at and interpreting the world that have also done a great disservice to the full richness of Creation, in many instances directly harming it. Such a view is limiting, much like a two-handed view of Kālī.

Thus, here in the realm of matter and our embodied selves is where the Dark Goddess lives and fights, calling upon us to resacralize nature. She is in the stuff of our bodies, the stuff of our world denied and rejected. She is in the dirt under our fingernails, in excrement, our blood, women's moon-time blood; she is in the juices of our love making, the burps of babies who come in all the colors of rain-washed Earth, the flesh of our beloved dead. She is the Divine immanent, here among us—and she is neither shy nor embarrassed about it. Rather, she shows up inebriated on the festival offered, drunk from the kissing of wounds.

Embodiment is also the power of the body's wisdom and intuition. It has correspondences in Earth and North. Her tool is a sacred diagram (such as a pentacle, labyrinth, yantra, or mandala) that provides a contemplative entryway to insight and deeper revelation of the inner mysteries. Engage practices such as breathwork, meditation, focusing, movement, yoga, and energy raising techniques to enter the arena of embodied wisdom. Draw on the power of your personal experiences to call up corporeal parables. These are the stories our bodies know that cross time and space; for example, childbirth, near-death experiences, and athletic episodes of being 'in the zone.'

Use resin incense and flowers to help awaken your senses and open the gateways to Her, with mindfulness to both external and internal realities. Find inside yourself the aromas, sounds, sensations, and feelings that arise, and pay particular attention to smells as activators of the subtle body. What you experience, this is an essence of the Dark Goddess.

ii. Relationality

Whether taken literally or metaphorically, the Dark Goddess is the one who shapes reality, creating and filling in spaces where we need to do the hard work of the spiritual journey. As we do this work, we may come to realize that relationships are our primary reminder of connection. We might not be surprised, therefore, to find the quality of relationality located at the Anahata Chakra, our heart center, a place of balance, potential, and equanimity. Seat of—or gateway to—the soul in many traditions, the heart is the place through which we experience the bonds of Creation: from the outside in and the inside out. This is the unstruck sound, a place of balance.

Simply put, relationality calls not on relationships themselves (it is not tied to any specific formulation or object of relationship). Rather, it calls on the urge toward and into connection. Thus, She catalyzes and invites exploration of all relationships and brings mindfulness of interconnections and interconnectedness to our attention. And this is particularly true when the relationships are ones not sanctioned by the orthodoxies of culture, religion, or family (among other frameworks). It is also true when She invites us in wholly new and different ways into explorations of relationships that we know and love.

Underscoring the quality of relationality is the fact that the Dark Goddess can take on roles that force us to move beyond limited conceptions of being-

T h e Dark Goddess embraces all; and she is a real and meaningful participant in the unfolding story of human evolution.

ness. For example, she can take on stereotypical masculine qualities and associated roles (such as warrior, hunter, sorcerer, and hero), while not allowing a relationship to a male to define Her;

she retains her life-giving abilities as She battles. Similarly, She can take on roles within the Maid-Mother-Crone trinity, each aspect of Goddess based on the particulars of female biology and a woman's life cycle, while not being tied to limiting definitions of what it means to be a woman. She is the independent goddess, one unto herself, while mothering the whole of Creation. This is how, for Her devotees, She is Maa, or mother. This is how she is protector and nurturer as much as she is the one who challenges us to our fullest potentials, demanding that we re-imagine and remove internalized biases and assumptions from what it means to be mother or crone, woman or man.

The second of her qualities, relationality has correspondences with the direction East and the Air element. Here, She engages our hearts in the world, pushing us into deeper experiences through the connections we have as well as through the possibilities for connection created in each moment—and this implies a move toward interdependence and toward the corollaries of ahimsa (nonviolence) and alliance building.

Do you wish to know her? Her tool of the East is the mirror. In it we see revealed our most authentic nature and the heart of all truths reflected through Her. As she gazes back at us, we may delve into the questions—especially those that relate us to our physically, emotionally, intellectually, and culturally informed notions of who we are—that get at the meaning of our identity and existence. By revealing our souls more completely as we engage with Her, the Dark Goddess brings us into closer, deeper and more honest relationships. This is the very stuff of soul making, the very stuff of awakening spiritual maturity. Here, She offers us the power of an unconditional embrace.

Practices such as seva (selfless service) and other efforts that create happiness, deepen heartfelt feeling states, and cultivate compassion (our ability to be with suffering), ignite this quality of the Dark Goddess. Practices you might wish to explore include dream work, scrying, regression and forms of transpersonal therapy, journaling, walking meditation, and conscious loving. These are all excellent ways to come closer to the relationality of the Dark Goddess. On your altar, use things that you particularly enjoy touching, picking up, and holding. Spend time with the objects of your choosing, allowing the holding to rekindle memories and awaken mythic connections. Choose a variety of tactile delights to incorporate into your ritual environment, such as textured fabrics, wood, stone, and fur. Above all else, bring yourself into love through devotion. What you experience, this is an essence of the Dark Goddess.

iii. Chthonic

Tell them you are the offspring of slaves and your mother was a princess in darkness. (*For Each of You*; Lorde, 1973)

A-U-M. OM. Mantras are not static.

The Pavamana Mantra from the Upanishads offers this prayer: from untruth to truth, from darkness to light, from death to immortality, lead us. The word chthonic, from the Greek *khthonios* meaning "in the earth," denotes that which is of or relating to the underworld. This quality of Dark Goddess, the chthonic, shares resonance with this most powerful and purifying of mantras. Chant it aloud now and feel the movement of sound around and within you:

OM

Asatoma sad gamaya

Aum uh'-suh-toh-maa suhd' guh'-muh-yuh

Tamaso ma jyotirgamaya

Tuh'-muh-soh-maa jyoh'-teer guh'-muh-yuh

Mrityorma amritam gamaya

Mriht'-yohr-maa uhm'-rih-tuhm guh'-muh-yuh

To feel the reverberations of the chanting as one speaks, intones or even listens to these Sanskrit syllables, is to know the essence of the chthonic, for through this mantra we are shown the potential to shift the very fabric of our beingness toward our dreams, our fullest realization and our ultimate bliss. In walking the path of the chthonic and in heeding the words of this eternal prayer, we are celebrating from our original yearning. We can not do this, however, without journeying to the underworld, to the place of our soul's passage from energy to manifest reality.

This quality, also home to our psycho-spiritual shadow is, therefore, essential on the spiritual journey because it facilitates growth, renewal, insight, and the burning away of all that baggage we tend to drag around with us unthinkingly. It is important to remember that the process of coming to understand the chthonic is

The power of the Dark Goddess lies in her claim to what we ignore. Darkness, indeed, is a metaphor for the deeper truth of our uncomfortableness with the full spectrum of energies the Divine offers.

not about simply moving between opposites (from one side of a polarity to the other, such as from darkness to light) in any linear or constant fashion. Rather, it is about the spiraling of awareness as we undertake and experience the quest. It is about the process of allowing the emptiness that comes through not knowing, through surrendering consciously, and through removing our self-centeredness to come to the fore. It is about iterations of consciousness as we work to release our attachments and move into greater clarity about life and living.

This work is particularly relevant where the chthonic is held psychically: at the third chakra, the Manipura Chakra, which is the seat of our individual will. The chthonic is also the third of her qualities, with correspondences in South and Fire. A downward-pointing red triangle is this quality's yantra, symbol of both creation and destruction. We might laugh to think of the underworld and fire connected together (since we may automatically conjure up images of a Christian Hell), but although often hot and fierce, the fires of the Dark Goddess burn for reasons other than damnation. In fact, they burn for purification and with the elemental power of light—that which gives form to energy and liberates us from misconceptions about phenomena in the material world. This is so because light brings with it the ability to perceive the external realm, and with this, we become aware of that which lies outside of ourselves. From here, the discernment of form brings the development of ego and the assertion of the self. The realm of the chthonic is therefore also our connection to individual power and the sacred dance we share with the Divine in working to discover our true nature.

The tool of the chthonic is the flame, that which brings illumination to the shadows, helping us to see the real and recognize the unreal. Practices such as those that induce trance, bring on altered states of consciousness, and assist inner sight are all beneficial toward understanding the realm of the chthonic. For example, shamanic drumming, trataka (focused concentration on one point or object), repetitive chanting, kriya yoga, and yoga nidra (practices that cultivate deep aware sleep) are all aids in coming to know this quality and its ramifications on gross, subtle and causal levels. Mudras too can help one focus the transformational energies of inner and outer worlds to great effect.

In working with this quality of Dark Goddess, you may wish to light candles and practice candle magick. Enjoy the play of the light on your chosen ritual objects, and see if you can enter into meditation, eyes open, while focusing about four feet away from you. Explore movement and dance, opening yourself to the physical landscape and the special features and curves of your own body and face. Create mudras spontaneously. Play with colors, sparkles, and eye-catching items on your altar, manifesting beauty and sweet flows of energy all around you. What you experience, this is an essence of the Dark Goddess.

iv. Cyclicity

You are not Atlas carrying the world on your shoulder. It is good to remember that the planet is carrying you. (Shiva, 2002)

The power of time, the space between dark and light where the yearning pulls us into the future, the internal mechanisms that reveal the oak tree from the acorn, the strands of our DNA, the ebb and flow of tides. Cyclicity is the fourth of her qualities, with correspondences in West and Water. Its locus is the Svadhisthana Chakra, seat of the self located at the pelvic floor. Some of the mystery may be revealed when we notice that the taste of pure water is unlike anything we know; it is clear, without particular qualities of sour or sweet, pungent or salty. Yet it is utterly refreshing and deeply satisfying to have a good drink. This is the revelation of the Dark Goddess through cyclicity.

Her tool in this place of ancestral knowing and remembrance is a water pot, which taps into the cycles of the moon and seasons, helping us open to the dynamic tension of the universe and the pull of our karmas through life, death and rebirth. Practices such as gardening that bring us into communion with nature and allow us to literally taste our efforts (as planting a vegetable garden would), as well as those that symbolically allow us to taste the fruits of life, such as devotional prayer, cooking, poetry writing, and other bhakti (devotional) offerings, can aid us in understanding and deepening our connection with this quality of the Dark Goddess.

Bring spices, food and drink to your altar, on one level remembering and honoring your ancestors within the cycle of life, and on another remembering and honoring the seasons and what it takes to nourish living beings. Food and drink on your altar also provides a remembering and honoring of the work we do and must do to feed the spirit. Cyclicity offers the prasad (blessed food) of the Dark Goddess. Nourish yourself through the virtues of this quality on all levels, heeding the power of the eternal return. What you experience, this is an essence of the Dark Goddess.

v. Antinomian

The antinomian is that which sits outside the normative, the socially acceptable, the so-called rational. It is the quality of the Dark Goddess that makes us think about our assumptions, reveals our internalized oppressions, and forces us to look directly at reality without the blinders of cultural, religious, and social frameworks. The fifth of her qualities, with correspondences in Center/Circumference and Ether, the quality of the antinomian is resonant with the Vishuddha Chakra or throat center. This is the seat of Her unfolding into consciousness through sound, the place where both reality and our karma are created through spoken words.

With the Dark Goddess, we can not only discover and work with the transformational powers she possesses, but also find sanctuary in her wisdom, especially in moments of despair, hopelessness, overwhelm, sorrow, and loss.

This space encompasses both sound and echo—the arising from the void into vibration and then into engagement with our sense of hearing, with the phenomenal world and back again. Since all sound is carried on ether, practices such as mantra repetition, chanting, kirtan (call and response singing), and sounding in general (with voice or instruments) are important to help uncover this quality of the Dark Goddess.

On your altar, you may wish to keep a rattle, drum, bell, or other instrument that augments or accentuates your voice or the silence as you do ritual, chant or sing. To know this quality is to move sound, create vibrations, and make waves in the fabric of your being and the world around you. It is to show up as you fully, authentically are, naked to the impositions of family and society. In

---

2 Evam astu is Sanskrit for "so mote it be." It is pronounced, aa-vuhm' uhs'-too. You may certainly use any other form of concretization or sealing of energy that you wish here.

this way and in a literal sense, digambara (sky-clad) ceremony can be liberating.

Her tool is the murti, or image that you will imbue with life-force energy for worship. The murti can be any object you wish to represent Her. It can be natural, such as a stone or tree, hand-crafted, or purchased in a shop; it may be quiet and located purely inside one's own heart. Unlock her secrets in the anointing, decorating, loving and honoring of Her image, this receptacle for the rapture that is her. Delight in that which will be present through it. If you have selected or created an image with eyes, gaze upon that gateway with intensity of concentration and surrendered self. What you experience, this is an essence of the Dark Goddess.

3. Integralizing Her

With the Dark Goddess, we can not only discover and work with the transformational powers she possesses, but also find sanctuary in her wisdom, especially in moments of despair, hopelessness, overwhelm, sorrow, and loss. We can find reassurance during times of challenge and crisis because She holds this part of reality too—and holds it along with everything else in loving kindness. Now that the psychospiritual space is prepared and we have readied ourselves through meditation, contemplation, and other practices, the time is ripe to welcome in the Dark Goddess.

We begin with activation of Her qualities. Once we feel the power of these essences, we turn inward, to the place of the indwelling Divine. We open to the qualities within ourselves, utilizing a focus on breath to bring us more deeply into present awareness. Place your hands on your heart, left on top of right palm and breathe as though doing so through this chakra, opening your heart. In this place of stimulated alignment between the outer experience of the qualities and inner awareness of them, imagine the coming together of breath and the deep within. Allow yourself to feel the universe of the inner planes together with consciousness fully present here at the root of your connection to all that lies beyond ego attachment and agendas.

Now, keeping your internal focus at the heart or center of the murti, say, "Welcome, Goddess (*Name*). Evam Astu!"[2]

With this invocation and awakening through intentional touch of prana (life-force energy), She is now present and alive—activated within your consciousness. Honor Her living presence wholeheartedly, with reverence and respect. Know that by this effort and through Her grace, you, and therefore we, are also just a little less encumbered— just a little more free.

May Her blessings fill you always.

References

Baring, A. & Cashford, J. (1991). *The myth of the Goddess: Evolution of an image*. London: Arkana/Penguin Books.

Begg, E. (1985). *The cult of the Black Virgin*. London: Arkana/Penguin Books.

Galland, C. (1991). *Longing for darkness: Tara and the Black Madonna, a ten-year journey*. New York: Viking/Penguin.

Gustafson, F. (1990). *The Black Madonna*. Boston: Sigo Press.

Harman, W. P. (1989). *The sacred marriage of a Hindu goddess*. Bloomington, IN: Indiana University Press.

Lorde, A. (1973). *From a land where other people live*. Detroit, MI: Broadside Press.

Matthews, C. (1992). *Sophia, Goddess of Wisdom: The divine feminine from black goddess to world-woul*. London: Aquarian Press.

Shiva, V. (2002). Heroes for the green century. *Time*. http://www.time.com/time/2002/greencentury/heroies/index_shiva.html.

Woodman, M. & Dickson, E. (1996). *Dancing in the flames: The dark goddess in the transformation of Consciousness*. Boston: Shambhala Publications.

Woodman, M. (1985). *The pregnant virgin: A process of psychological transformation*. Toronto: Inner City Books.

Haiku Sequence

spiritus sanctus
I breathe in
the cat breathes out

temple of my heart
no gods   no goddesses...
sweet jasmine

around the tree roots
compost of a century
only inches deep

autumn saunter
returning home
on the rutted trail

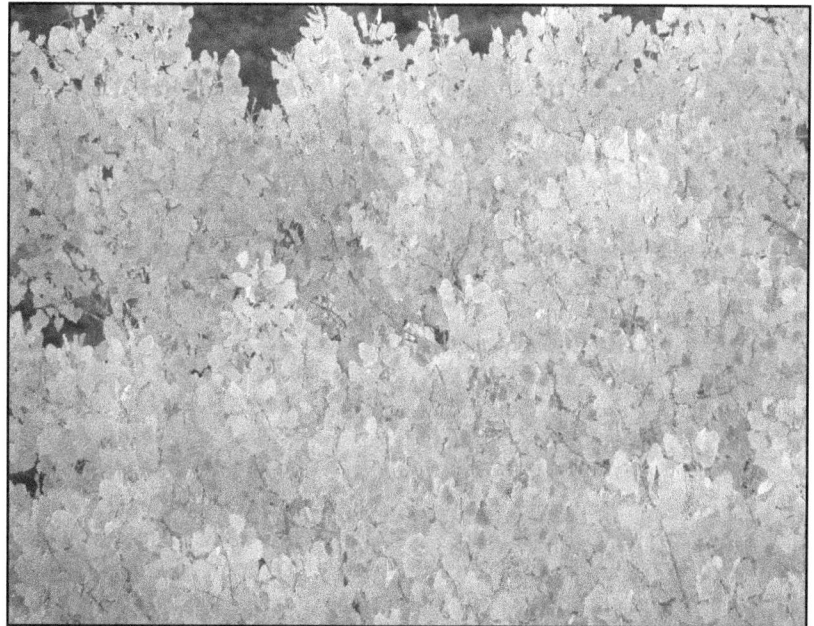

Photo: Jürgen Werner Kremer

winter solitude
above the darkened cabin
a moon and stars

Michael Sheffield

Transformative Travel

An Enjoyable Way to Foster Radical Change

Susan L. Ross

This article explores how travel, when approached in a conscious way, can be a widely available, individually tailored, and enjoyable way to gain self awareness, spiritual experience, and an expansion of consciousness. First, transformative travel is defined and described through a review of literature from a broad range of disciplines. Ways to maximize the transformative potential of a given travel experience are then identified, followed by a description of the main types of transformative travelers.

The topic of transformation during travel and travel-related conditions that produce personal transformation have escaped extensive scrutiny and attention. There are many studies that show or provide evidence of transformative travel, but few studies have examined elements of travel that contribute to transformative experience. In order for travel to serve as useful means for trans-

Susan L. Ross, Ph.D., teaches Recreation Therapy and specializes in post traumatic stress and adventure therapy. She is author of the forthcoming book, *The Hidden Half of Transformation: Nine Stages that Integration Peak Experiences into Daily Life*. Ross co-leads transformative travel to sacred sites to cultivate ancient ritual, relationship with living energy, and partnership with elders. Contact: San Jose State University, One Washington Square, 0021 Hospitality, Recreation and Tourism Department, Spartan Complex, 52, San Jose, CA 95192, sross@casa.sjsu.edu; cell: 831-295-2168; fax: 408-924-3061.

formation of consciousness, the conditions that contribute to transformative experiences must be identified.

Defining Transformative Travel

Transformative travel involves two main concepts: transformation and travel. Different disciplines define personal transformation in various ways depending on contexts, worldviews, and purposes. A personal transformation is defined here as: a dynamic sociocultural and uniquely individual process that (a) begins with a disorienting dilemma and involves choice, healing, and experience(s) of expanding consciousness towards the divine; (b) initiates a permanent change in identity structures through cognitive, psychological, physiological, affective, or spiritual experiences; and (c) renders a sustained shift in the form of one's thinking, doing, believing, or sensing due to the novelty of the intersection between the experiencer, the experience, and the experiencer's location in time. Here, the terms transformation and personal transformation are used interchangeably.

Transformation has long been asso-

ciated with travel but the concept of transformative travel is a recently used term. Although culture shock theory did introduce the potential for the conscious use of cross-cultural experiences as a means for "self-development and personal growth" towards "higher levels of personality development" (Adler, 1975, p. 14), it was Jeffrey Kottler (1997) who first introduced the term transformative travel into scholarly discourse. Kottler helped countless clients—including himself—engage in therapeutic transformative travel and claimed that there

The action of travel itself has been identified as a modern rite of passage.

is "no other human activity that has greater potential to alter your perceptions or the ways you choose your life" (Kottler, 1998, p. 14).

Some tourists intentionally seek life-changing experiences through travel. People are motivated to transform through travel for a variety of reasons and through a range of media. Various travelers seek transformation through intellectual or physical challenges, while others seek a transformation of the heart or spirit through travel that allows them to be creative or altruistic—and yet others seek to deepen their feelings of connection to a deity, culture, ecosystem, or place (Kottler, 1997, 1998, 2002). Conceptually, transformative travel provides a framework to examine those sojourners who, regardless of secular and sacred motivations, locations, or activities, wish to expand consciousness through radical personal change.

Although travel can change tourists' lives, few authors have actually used the term transformative travel to denote this type of travel behavior. Through psychological lenses, transformative travel has been defined as a process that involves the actualization of "something missing" driven by "intellectual curiosity, emotional need, or physical challenge" (Kottler, 1998, p. 26). Transformative travel can also be the "result of a process that begins with some type of experience that does not fit within the boundaries of the traveler's assumptions, expectations, worldviews, or cultural paradigms" (Robertson, 2002, p. 4). The use of the term transformative travel as opposed to transformative tourism directly concerns the traveler's perspective and experience, as opposed to the industry that provides goods and services to the traveler.

Transformative travel is defined here as sustainable travel embarked upon by the traveler for the primary and intentional purpose of creating conditions conducive for one or more fundamental structures of the self to transform. This definition of transformative travel includes a key distinguishing element: the traveler's conscious intention. None of the authors who have used the term transformative travel (Kottler, 2002; Kottler & Montgomery, 2000; Lean, 2005; Robertson, 2002) distinguished the role, if any, of the traveler's intention to transform. Of the relevant studies, nearly all attempted to either discern the elements of the travel that helped people transform or to discern the ways in which the sojourner transformed, with the vast majority focusing upon the latter. The literature reflects analyses of situations where individuals transformed during travel—regardless of their intentionality. Sometimes travel happens to be transformative for a given

individual, but for the definition used here, the transformation must be intentional on the part of the traveler for the experience to be considered transformative travel.

Transformative travel includes a list of concrete actions that, based upon lived experience and literature, might create conditions conducive to fostering transformative experiences while traveling. The following is a noninclusive and nonprescriptive list of transformative travel activities:

1. Dwelling at sacred sites.

2. Engaging in ritual and ceremony.

3. Initiating regular group sharing sessions.

4. Being in nature and connecting to natural sites through our body/mind/heart.

Transformative travel and transformative tourism aim to honor the delicate interplay between the self and anyone who is different, or the 'other,' during travel.

5. Spending money with a sustainable ethic.

6. Learning esoteric and common history.

7. Engaging in multiple means of self-exploration such as: reflection, yoga, expressive art, group activities or exercises, journaling, nature hikes, or guided meditations, to name a few.

8. Talking to, listening to, and learning from willing indigenous teachers, shamans, community members, and children.

9. Engaging in physically challenging activities and/or adventure.

10. Giving and providing services to families, children, and communities (Ross, 2005, p. 54).

In particular, in order to avoid offending or exploiting host country residents, transformative travel and transformative tourism aim to honor the delicate

interplay between the self and anyone who is different, or the 'other', during travel. Transformative tourism and travel include responsible travel and sustainable practices guided by values of respect for both the host peoples and all ecosystems. It is critical to attempt to acknowledge the presence of differences in privilege. Travel with intention to transform needs also to involve choices that attempt to contribute to the flourishing of all life. This kind of ethic involves being a consumer of sustainable enterprises and products that explicitly strive to honor and give to the earth, the local culture, and individuals.

Maximizing the Potential for Transformation

It has been established that novelty is a critical precursor to transformation (Fosha, 2006) and it is clear that novelty is inherent to travel (Kottler, 1998; Williams & Soutar, 2009) and travel motivation. During travel the sojourner separates from usual influences, pressures, and structures, inducing an inner readiness and creating conditions conducive for transformation. Kottler (1998) suggests that travel can serve to remove individuals from typical unhealthy patterns, allowing them to act more freely, to experiment with new ways of being or lifestyles, to escape, or to seek solutions when life seems to be unraveling. Travelers can internally prompt a personal transformation by creating a mindset that makes a person ripe for change.

A pilot study (Robertson, 2002) examined the transformative elements of travel in a group of experienced travelers over the age of 60. The results indicated that transformation during travel is supported through, (a) perception of the similarity and contrast between host and home culture, (b) effective tour guides, (c) internal and external changes in meaning schemes (i.e., that which determines how and what we see), (d) the quality of travel planning or preparation, and (e) interactions with local people. Interactions with host people seem to leave a "marked impression, even if these interactions were very brief" (p. 10). Intercultural experiences can, for instance, place travelers in "a position from which it becomes possible to see the inadequacies of [one's] own society more sharply" (Turner & Ash, 1976, p. 49). It is this contrast which can prompt drastic personal change.

A web-based study of 62 individuals from 17 nations who had extended stays in foreign countries (Lean, 2005) delineated conditions that lead to personal transformation through travel. The factors found to contribute to transformation included experiences of intercultural immersion, the length of stay, and the person's openness to change. Four main factors led to greater likelihood for transformation and positive behavioral change resulting from travel: (a) a destination that pulls the individual as far as possible from known experiences (novelty), (b) intimate intercultural experiences involving in-depth discussions, (c) activities that stimulate contemplation resulting in meaning-making of the traveler's experiences, and (d) post-travel activities that help the sojourner to continue to reflect upon and extract meaning from the travel.

One group of students who had been back in the United States for one year after a study-abroad program participated in a qualitative research study that used in-depth interviews (Tuffo, 1994). The data revealed several factors that contributed to the students' (transformative) learning: (a) significant relationships that facilitate cross-cultural dialogue and emotional support, (b) pivotal incidents and crises that challenge the students ontologically, and (c) critical reflection (resulting in a transformation of meaning perspective).

Finally, a study that examined four unique populations of travelers (Kennedy, 1994) elucidated the ways in which sojourning contributes to transformative travel experiences. The four travel groups included students and adult travelers, business travelers, foreign sojourners to the United States, and long-term sojourners in the Peace Corps. The study identified factors that influence or do not

influence personal transformation during travel. Personality type and learning style did not determine whether one had a transformation, nor did the reason for travel or length of stay. The degree of ego development did determine both the level of learning and the depth of the transformational learning experience.

In addition to these empirical studies, there are two factors that, by virtue of

Travel can serve to remove individuals from typical unhealthy patterns, allowing them to act more freely, to experiment with new ways of being.

being key transformative elements in lived experience and literature, could play a role in transformation through travel. These are movement and the nature of one's travel companions. The seminal work on ritual completed by anthropologist Victor Turner (1969) unveiled the deep significance of movement. When individuals and communities embark on a journey and leave behind the personal and what is known, sociocultural rules or structures are suspended, providing a certain experience of freedom from restraints of structure—this 'antistructure' issues freedom to be in ways not possible during ordinary daily life.

Experiences of antistructure encourage previously unavailable possibilities of relationship that foster a special quality and role of community Turner (1969) calls communitas, which accentuates the bonds of being human in a "communion of equal individuals" (p. 96) and relieves pilgrims of socially sanctioned roles and hierarchical division. The results of an informal survey by this author showed several students claimed the most transformative element of the travel was their experiences of relationship they described in ways reflective of Turner's communitas. This result was confirmed in a recent survey of transformative (mystic) travelers (Nasiri, 2009).

Anecdotal observations supported by the literature (Arnould & Price, 1993;

Williams & Soutar, 2009) illustrate the plausibility that the quality and nature of one's travel companions have an impact on the potential for transformation to occur. Groups have character that is difficult to predict and can be influenced by just one strong personality. A highly committed, devoted, flexible, mindful, and yet playful group, for instance, will hold different possibilities than a group of people containing a few individuals who are inflexible, prone to anger, or highly dramatic. Some people travel with family members, which also can deeply influence the potential for transformation to occur. As a corollary to the effect of travel companions, some sojourners travel alone—a means of travel that offers particular possibilities for transformation, such as time spent in reflection, contemplation, and meaning-making, not to mention an enhanced readiness for diverse and intercultural experiences.

In summary, literature and some anecdotal evidence indicate the following conditions foster the potential for a transformative experience to occur through travel: Traveler intent; Quality preparation; Place and space; Novelty; Movement; Intercultural experiences; Spiritual or therapeutic activities that stimulate personal growth, a sense of the sacred, contemplation, meaning-making, and sharing; Quality and nature of guides; Consciousness and awareness during travel; Extent of one's ego development and inner readiness; Travel companions and social value; Post-travel integration activities.

Understanding Transformative Travel and the Transformative Traveler

Transformative experiences are individually unique and can happen anywhere and at any time. Because travel is particularly conducive to transformation, any travel can theoretically be designed and experienced so as to create conditions for profound personal change. In a review of literature, however, there emerged trends in the types of travel where intentionality, conscious choice to evolve, and transformation were evidenced or implied. The types of travel that tend to be ripe for transformation to occur include: rites of passage, study abroad, heritage travel, pilgrimage, adventure/challenge, service learning, healing or therapy, and spiritual travel.

In an earlier study, I combined the themes found in the literature to distill six distinct categories of transformative traveler: pilgrims (traditional and goddess), mystic travelers, diaspora sojourners, initiates participating in travel as a rite of passage, secular travelers seeking healing and personal growth, and learners intending to gain consciousness through knowledge (Ross, 2008). In the next section, each of these transformative travelers is discussed in greater detail.

The Pilgrim Traveler

Pilgrimage has "become widespread and popularized in recent decades" (Olsen & Timothy, 2006, p. 1), and is a resurging segment of international tourism, whether sacred/secular, traditional, or contemporary. Pilgrimage is an ancient phenomenon dating back many

Traditional or historical pilgrimage.

Travelers who respond to a religious commitment embark upon pilgrimage in which the pilgrim seeks to be affected by both the journey and the spiritual locale or center (Turner, 1973). A narrative analysis of 121 modern Christian (predominantly female Anglo-American) pilgrims yielded specific ways in which pilgrims might change or transform (Presnell, 2001). Those participating in pilgrimage deepened discipleship, increased service to others, or began to serve as advocates towards alleviating political tensions. One participant implied the transformation in his description: "We make our journeys and they make us" (p. 75).

Sacred sites (place and space).

During pilgrimage (and new age or mystic travel) "people travel and submit themselves to certain environments in order to make something happen within themselves" (Osterrieth, 1985, p. iv), thus highlighting the significant role of place to transformative experience.

such as in places called tirthas (sacred fords or crossings), in dhams (divine abodes), or in a pitha (benches or seats of the divine). Sacred places can also be called "transformational sites" (Osterrieth, 1985, p. 99).

Once at such a sacred site, dwelling there can be transformative. There is a link between how dwelling and journeying as elemental "modes of being-in-the-world" can "give meaning to space, place, and movement" (Osterrieth, 1985, p. 60). Data indicate that dwelling in liminal spaces can be transformative, as evidenced by accounts where "the pilgrim may experience things during the journey that do not occur at home 'miracles' may occur in which the veil between self and other, the profane and the sacred" is lifted (p. 33).

Goddess pilgrimage.

Although this topic could be included in the next subsection on new age tourism, goddess pilgrimages include some of the qualities of traditional pilgrimage, such as the significance of communitas,

The types of travel that tend to be ripe for transformation to occur include: rites of passage, study abroad, heritage travel, pilgrimage, adventure/challenge, service learning, healing or therapy, and spiritual travel.

millennia prior to the development of tourism and is often referred to as one of the historical roots of modern-day travel and the tourism industry (Schmidt, 1980). Historically, religious travel has involved personal (Aziz, 1987) or communal travel or movement that can be repeated with relatively little change, to a geographical location known to be sacred (Turner & Turner, 1978) for the primary purpose of following religious devotion and communion with the divine. Pilgrimage is strongly aligned with intentional transformative travel. Three relevant themes address the ways in which pilgrimage is transformative: traditional religious pilgrimage, sacred sites (place and space), and goddess pilgrimage.

The pilgrim often travels to an "actual place containing the actual objects of the past, whose very stones seem to emit the never obliterated power of the first event" (Turner & Turner, 1978, p. xv). An important characteristic of sacred sites is that the locations offer "distinctive features" that are opposite of one's "ordinary everyday life." (Osterrieth, 1985, p. 93).

The prevailing East Indian culture (among many others), for instance, has embedded within the language an ontology that sustains the notion that place/space can hold and perhaps even emit transformative spiritual power. One pilgrim in India wrote an essay of her experience explaining that there exist 'powerful places' where the pilgrim or 'sacred sightseer' can receive teachings,

the worship of a deity, and the increasing sacralization of the journey as pilgrims move closer to the shrine or sacred centre (Turner & Turner, 1978). Goddess pilgrims (and mystic pilgrims) often perceive this travel as an act of religious identity and political consciousness, where personal and social transformation is ontologically the same and integral to the journey. Some pilgrims express their expectation that the journey will be transformative before embarking upon the journey. One pilgrim said, "I know intuitively that what happens over the next two weeks is going to transform me forever. She [Goddess Durga] is calling me" (Chamberlain, 2001, p. 70). Chamberlain quoted from this participant's reflective journal upon return, where she described transforming her conditioning after having witnessed rituals involving sacrifice:

I was able to hold a multiplicity of emotions and possible interpretations. By engaging with the subject in a way that was open to understanding the role of the sacrifice and its necessity in upholding the ritual I was able to see it in a new way. (p. 90)

Goddess pilgrimage also provides women a unique opportunity to heal through reclamation of the feminine that renders a transformation in identity or enhances a felt sense of self. In this way, pilgrimage can result in a "radical re-inscription of the female body" impelled by experiencing ancient images of the feminine and a rich context to "re-imagine and re-experience [the feminine] through symbolic activity and ritual" (Rountree, 2002, p. 475).

The Mystic Traveler.

The 'New Age' has invited a quality of spirituality that has widened the scope of traditional pilgrimage, and supports research indicating leisure activity can be a means for self-actualization or transcendence. Anthropologists Victor and Edith Turner (1978) acknowledge three types of modern pilgrimage including one termed New Age (here termed the mystic tourist). They provide an important distinction between the pilgrim and the mystic tourist: "Pilgrimage may be thought of as extroverted mysticism, just as mysticism is introverted pilgrimage. The pilgrim physically traverses a mystical way; the mystic sets forth on an interior spiritual pilgrimage" (p. 33). The mystic tourist travels with attention to both an inner and outer journey, possibly even viewing the two as mirrors of one another. The concept of the mystic traveler is probably the most closely aligned with transformative travel as defined here, as mystic travel indicates a traveler intending to transform spiritually.

Mystic or esoteric tourists are travelers who "seek personal enlightenment" (Arellano, 2006, p. 11) through teachings from local people or even "teachings from 'another dimension'" (p. 11). Mystic tourists sometimes intend to achieve reflexive awareness of their bodies, or to have embodied and even sensuous experiences meant to re-live ancient ways through bodily engagement. In so doing,

the experience is "aesthetic, kinesthetic but also self-transformational and healing" (p. 11). One such traveler, with intentions to merge realities on his hike on the Inca Trail to the archeological ruins of Machu Picchu in Peru, shared his intentions, saying "I want to enter the city [Machu Picchu] as the original inhabitants did. It was more important to me that my steps echoed the Incas" (p. 9).

Travel in this way can also involve engaging in activities that invite thoughtfulness, reverence, reflection, sharing, or nonordinary states of consciousness such as meditation, ritual, ceremony, and even hallucinogenic plants. Ceremony and ritual in the context of a wilderness-based host community can lead participants to experience unity or oneness and to transcend embedded and unwanted North American and European conditioning of individualism and separatism.

To travel is to engage in privileged activity, especially since airplane travel is so expensive. Conversely, for the mystic tourist, a sacred site of pilgrimage could be anywhere, because individuals (rather than society or culture) define the center of sanctity. The center is intimately personal and made holy through private meaning-making. A depth psychologist, senior minister, and devoted follower of depth psychologist Carl Jung's work, wrote an essay about his "pilgrimage" to Jung's retreat home in Bollingen Switzerland. In a reflection of his mystical experience, he shared:

No one was there at the time. I stood and stared, trying to believe I was really there. Jung built this retreat as a physical representation of his own wrestling with his soul, with Psyche. The buildings, and the carvings he made, are his expressions of his own inner journey. It was the closest I could come to experiencing Jung's presence. For years, I had been reading his writings, experiencing his intellect, but then, at Bollingen, I could feel something of the man still imbued in the stone and wood. (Budd, 1989, p. 121)

In this journey, Budd gained a critical experience coupled with an epiphany as shown when he explained, "I realized I was meeting my self" (p. 122).

The Diaspora Traveler

The concept of 'Diaspora' emerged as a term identifying Jewish people dis-

placed from their homeland. It has since widened in scope and is now used to describe any people who have left their homeland, either by choice or otherwise. Diaspora tourism, also called personal heritage tourism (Timothy, 1997, 2001), roots tourism, or genealogy tourism, refers to "restless or halfway" populations who sojourn back to their home countries to engage in the re-making and de-making of their identities. In their home countries, such pilgrims can relive a mythic past.

Diaspora tourism has many qualities similar to New Age travel, but not all diaspora tourists are mystic tourists. Although diaspora tourists are motivated "to embark upon a journey of self-change," the motivation is seeded in a desire to satiate "feelings of alienation and fragmentation" (Cohen, 2006, p. 79) as related to cultural and biological roots and "to transcend the present and engage with the past" (Cameron & Gatewood, 2003, p. 1). The spiritual qualities of this kind of tourism are reflective of pilgrimage because individuals set out on an existential or spiritual quest for meaning and identity resolution.

The Initiate Traveler

Some travelers who appear to travel for the intended purpose of evoking transformation do so by embarking upon a rite of passage or initiation ceremony. This type of travel typically involves an intention to acquire one or more of the following: knowledge, healing, spiritual experience, change in status/purpose, societal liberation, or personal transformation. The main venues through which this rite of passage might occur include: wilderness/nature activities, adventure therapy, traditional and contemporary initiation or rites of passage, and adventure challenge such as mountaineering or adventure programs.

A preeminent scholar of world religions, rites, and symbols, Mircea Eliade (1958/1994) states that initiations "represent one of the most significant spiritual phenomena in the history of humanity" (p. 3). Many initiation ceremonies and rites of passage are means through which the culture can sustain meaning and values, and re-create itself. The action of travel itself has been identified as a modern rite of passage because it is reflective of van Gennep's and V. Turner's ancient rite of passage processes that include separation, liminal, and incorporation phases (Inkson & Myers, 2003).

Rites of passage or initiation serve as central functions in individual and societal health, even in modern and post-modern times. Malidoma SomÈ (1993), a traditionally initiated West African Dagara medicine person and scholar, suggests that the prevailing culture in the United States is lacking in healthy and whole rites and initiations. Elizabeth Cogburn, a pioneer of developing modern rituals and ceremonies, purports that sacred ceremony is "humanity's most highly developed meta-system for the conscious transformation of energy" (Chapin, 1984, p. 42).

It should be noted that this type of activity can risk defamation of the cultures from whence the symbolic structures came, or worse, spur untrained replication of ancient practices. Herwood (1994) explains that the "surface of a ceremony hides lengthy and rigorous preparation not visible to outsiders" (p. 12). He goes on to say that indigenous ceremony, such as rites of passage, is steeped in "sacred mythology and cosmology" that is resultant of countless generations (p. 12).

An international program based in Canada called "Rediscovery" facilitates rites of passage for youth of diverse heritage overseen by indigenous elders. A central component of this wilderness program for youth aims to help them to recover their own indigenous heritage and to foster the "making of whole human beings" (Lertzman, 2002, p. 30). Contemporary versions of pan-cultural vision quest ceremony describe the process as a means to tap into archetypal forms of consciousness (Suler, 1990). This type of initiation includes dwelling or wandering, relating self to world, and attaining insights via the appearance of signs. All 20 participants in Jensen's study (2004) reported experiencing personal transformation. More specifically, data showed that most of the experiences of transformation involved a profound shift in the traveler's anthropocentric worldview of separateness from nature. Respondents reported a new feeling of interconnectedness and ecological consciousness. This might be called an epochal transformation or embodied transformative learning.

The Secular Traveler

Travel has the potential to be a viable and intentional therapeutic approach to personal growth and healing. Healing-

> Goddess pilgrimage also provides women a unique opportunity to heal through reclamation of the feminine that renders a transformation in identity or enhances a felt sense of self.

motivated transformative travel overlaps many of the other categories discussed here (i.e., heritage tourism, rite of passage, pilgrimage, and adventure therapy). For example, a wilderness journey, either facilitated by a guide or self-facilitated, might be pursued for the purpose of healing. It is possible that "solitude, isolation, or novel environments in themselves are not enough; [one] must also complete tasks that are therapeutic and educational" if one wants transformation and lasting healing (Kottler, 1998, p. 25).

Potentially therapeutic structures in transformative travel occur when the traveler (consciously or unconsciously) pretends to live in a different time or culture, escaping daily life to gain perspective upon life, or experimenting with different social roles/behaviors. When travel is adeptly self-facilitated for personal growth, a week in Aruba is "worth three years in therapy" (Kottler, 1998, p. 24). Based upon clinical observation of his patients and his own travel,

Kottler suggests ways in which people who want to heal and transform through travel might self-facilitate transformative travel. Sojourners wanting personal change, for instance, ought to make public commitments prior to travel that identify the growth-producing action the traveler will undertake, such as sky diving or staying in a host-family home rather than a Westernized hotel. Kottler suggests experimenting with new ways such as doing "the opposite" of what one might ordinarily do, and processing experiences by journal writing or sharing with others. Lastly, upon returning from travel individuals ought to make changes to demonstrate the newly found "self" or awareness, such as quitting a dead-end job or ending an unhealthy relationship. Importantly, Kottler qualifies the "processes that lead to self-transformation" as including experiences before, during, and after travel.

Some adventure travel specialists facilitate personal growth. Of the hundreds of studies completed in this area, one inquiry specifically points to the need for research that includes the integration of transformative travel. Cushing (1999) set out to examine the types of transformations produced through a classical 'Outward Bound' adventure course geared to develop personal growth and outdoor skills, using long ethnographic interviews with 22 students one month prior to the course, on the last day of travel, and three months post-course. The participants reported that upon return home, it was substantially harder to maintain personal changes made during the travel than they had imagined it would be. The analysis indicates that while participants might have cherished transformative experiences, implementing and living the change had many obstacles.

Another avenue by which individuals hire professional travel guides for the purpose of healing and growth is adventure therapy, which involves traveling via an array of formats such as contrived adventure challenge courses, initiations, wilderness trips, or a combination. The purpose of these experiences can range from personal growth to clinical treatment. Participants experientially engage personal and interpersonal dynamics of trust, challenge, risk, cooperation, communication, courage, assertiveness, and appreciation of nature. Adventure therapy tends to involve disorienting dilemmas, challenge, risk, group or individual emotional processing, heightened senses, and nature experiences. The major distinction of adventure therapy is that the growth and healing is facilitated, while most travel is not (unless the individual has capacities to self-facilitate or hires a company/individual to do so). Adventure therapy excursions are a form of transformative travel because a primary purpose for the individual's excursion is healing and personal growth.

The Learner Traveler

A few studies have examined people who travel to learn, and in doing so,

Some people are using transformative travel as a way to radically transform consciousness in a short amount of time and in a way that is pleasurable.

transform. For example, research completed on a faculty-led study abroad to Cuba discerned the degree and kinds of transformative learning that can occur in this context (Brown & Smith, 2003). During travel, the coresearchers engaged in a purposeful and diverse learning community of 30 participants using transformative learning theory and practices. The data included interviews that transpired 11 months post-travel and narrative data from academic reflective papers. Results showed a perspective transformation about the host people and

about government program effectiveness toward social reform. All participants claimed that the intercultural experiences with hosts produced the most meaningful transformation, which corroborates with other research findings that also report the transformative effects of intercultural experiences (Lean, 2005; Robertson 2002).

Transformative education theory as presented by Boyd and Myers (1988) discusses the key role travel can play in opening intrapsychic dialogue between one's ego and "other aspects of the self" (p. 267). An empirical study of sojourners seeking transformative learning through travel found that ego development does "determine the level of learning and the depth of one's transformational experience" (Kennedy, 1994, p. iii). In contrast, neither personality type, learning style, length of stay, nor reason for the travel affected the likelihood of having a transformative travel experience.

Transformative Travel and Global Change

As the Indian philosopher and yogi Sri Aurobindo (2003) foretold, "If humanity is to survive, a radical transformation of human nature is indispensable" (p. ix). Even though radical transformation might be something many seek, Aurobindo argues that "only spiritual realization and experience can achieve the change" (p. xi) our current situation requires. As individuals, many have become increasingly aware that "without an inner change man [sic] can no longer cope" (p. 33) with the ominous demands made by the institutions we created with the linear mind and ego. As a collective we have reached a point of necessity, an "evolutionary crisis" (p. ix), which asks us to consciously grow not only for ourselves, but for the collective.

In response to this urgency, some people are using transformative travel as a way to radically transform consciousness in a short amount of time and in a way that is pleasurable. It is arguable that there is any other single activity that so thoroughly exposes us to difference on every level of human interaction. It is, after all, exposure to difference—that which is beyond lived experience and even capacities—that spurs consciousness to stretch beyond stagnant and even

hardened patterns, and to become transformed anew. With a shorter timeframe than most types of psychotherapy, and a level of saturation and experiential wholeness that could rival even the best of spiritual retreats, travel could potentially provide a rich and unique opportunity for transformation of consciousness.

References

Adler, P. (1975). The transitional experience: An alternative view of culture shock. *Journal of Humanistic Psychology, 15*(4), 13–23.

Arellano, A. (2006, March). Reconfiguring the roots/routes of the sacred: The tourists' quest for the lost vagabond. Conference session presented at the Travel and Tourism Research Association conference, Belfast, Northern Ireland.

Arnould, E. J., & Price, L. L. (1993). River magic: Extraordinary experience and the extended service encounter. *Journal of Consumer Research, 20*(1), 24-45.

Aziz, B. N. (1987). Personal dimensions of the sacred journey: What pilgrims say. *Religious Studies, 23*, 247–261.

Boyd, R. D., & Myers, J. G. (1988). Transformative education. International *Journal of Lifelong Education, 7*, 261–284.

Brown, P. C., & Smith, C. A. (2003, October). Travel as transformation? A Cuban experience in education. In *Proceedings of the Fifth International Transformative Learning Conference, Transformative Learning in Action: Building Bridges Across Contexts and Disciplines*. New York: Teachers College, Columbia University.

Budd, D. E. (1989). Pilgrimage and enlightenment. *Journal of Religion and Health, 28*, 119–127.

Cameron, C. M., & Gatewood, J. B. (2003). Seeking numinous experiences in the unremembered past. *Ethnology: An International Journal of Cultural and Social Anthropology, 42*(1), 55–71.

Chamberlain, L. K. (2001). Embodying the goddess Durga: A pilgrimage to the mother goddess of paradox. Unpublished master's thesis, California Institute of Integral Studies, San Francisco.

Chapin, R. (1984, Spring). Warriors of the beauty way: Realizing the power and possibility of human potential by creating in beauty, an interview with Elizabeth Cogburn by Ross Chapin [Electronic version]. *In Context: A Quarterly of Humane Sustainable Culture, 5*, 42–46.

Cohen, E. H. (2006). Religious tourism as an educational experience. In D. J. Timothy & D. H. Olsen (Eds.), *Tourism, religion and spiritual journeys* (pp. 78–93). New York: Routledge.

Cushing, P. J. (1999). Translating transformation into something real. *Pathways: The Ontario Journal of Outdoor Education, 12*(1), 26–28.

Eliade, M. (1994). *Rites and symbols of initiation: The mysteries of birth and rebirth* (W. R. Trask, Trans.). Putnam, CT: Spring. (Original work published 1958)

Fosha, D. (2006). Quantum transformation in trauma and treatment: Traversing the crisis of healing change. *Journal of Clinical Psychology, 62*(5), 569–583.

Herwood, B. (1994). The ceremonial elements of non-native cultures. *The Journal of Experiential Education, 17*(1), 12–15.

Inkson, K., & Myers, B. (2003). "The big O.E.": International travel and career development. *Career Development International, 8,* 170–181.

Jensen, A. J. (2004). Earth medicine: Transformation in nature. *Dissertation Abstracts International, 65*(3), 1578.

Kaye, R. (2006). Soul of the wilderness, the spiritual dimension of wilderness: A secular approach for resource agencies. *International Journal of Wilderness, 12*(3), 4–7.

Kennedy, J. G. (1994). The individual's transformational learning experience as a cross-cultural sojourner: Descriptive models. Unpublished doctoral dissertation, The Fielding Institute, Santa Barbara, CA.

Kottler, J. A. (1997). *Travel that can change your life: How to create a transformative experience.* San Francisco: Jossey-Bass.

Kottler, J. A. (1998). Transformative travel. *The Futurist, 32*(3), 24–29.

Kottler, J. A. (2002). Transformative travel: International counseling in action. *International Journal for the Advancement of Counseling, 24,* 207–210.

Kottler, J. A., & Montgomery, M. (2000). Prescriptive travel and adventure-based activities as an adjunct to counseling. *Guidance and Counseling, 15*(2), 8–11.

Lean, G. L. (2005). Transformative travel and the creation of sustainability ambassadors: Literature review. Unpublished Bachelors project, University of Western Sydney, Australia.

Lertzman, D. A. (2002). Rediscovering rites of passage: education, transformation, and the transition to sustainability. *Conservation Ecology, 5*(2), 30.

Nasiri, N. (2009). An Analysis of Youth Motivations for Spiritual Travel and Initiation in the Andes: A pilot study. Unpublished master's project. San JosÈ State University, San JosÈ, CA.

Olsen, D. H., & Timothy, D. J. (2006). *Tourism, religion and spiritual journeys.* New York: Routledge.

Osterrieth, J. A. (1985). Space, place and movement: The quest for self in the world. Unpublished doctoral dissertation, University of Washington.

Presnell, W. B. (2001). The power of a pilgrimage to the holy land. Unpublished doctoral dissertation, Drew University, Madison, NJ.

Robertson, D. N., Jr. (2002). Modern day explorers—the way to a wider world [Electronic version]. *World Leisure, 3,* 35–42.

Ross, S. (2005). Transforming TR through Travel. Poster session presented at the Therapeutic Recreation Educator Conference, Itasca, IL.

Ross, S. L. (2008). The integration of transformation: A cooperative inquiry among women after transformative travel. *Dissertations Abstracts International, 69* (3).

Rountree, K. (2002). Goddess pilgrims as tourists: Inscribing the body through sacred travel. *Sociology of Religion, 63*(4), 475–497.

Schmidt, J. C. (1980). Tourism: Sacred sites, secular seers. Unpublished doctoral dissertation, State University of New York at Stony Brook.

SomÈ, M. (1993). *Ritual, power, healing, and community.* New York: Penguin Press.

Sri Aurobindo (2003/1963). *Future evolution of man: The divine life upon earth* (2nd ed.; P. B. Saint-Hilaire, Comp.). Twin Lakes, WI: Lotus.

Suler, J. R. (1990). Wandering in the search of a sign: A contemporary version of the vision quest. *Journal of Humanistic Psychology, 30*(2), 73–88.

Timothy, D. J. (1997). Tourism and the personal heritage experience. *Annals of Tourism Research, 34,* 751–754.

Timothy, D. J. (2001). Sacred journeys: Religious heritage and tourism. *Tourism Recreation Research, 27*(2), 3–6.

Timothy, D. J., & Olsen, D. H. (2006). *Tourism, religion & spiritual journeys.* New York: Routledge..

Tuffo, K. M. (1994). Narratives of student experience, reflection, and transformation in experiential cross-cultural learning. Unpublished masters thesis, University of British Columbia, Canada.

Turner, V. (1969). *The ritual process: Structure and anti-structure.* Ithica, NY: Cornell University Press.

Turner, V. (1973). The center out there: Pilgrim's goal. *History of Religion, 12,* 191–230.

Turner, V., & Turner, E. L. B. (1978). *Image and pilgrimage in Christian culture.* New York: Columbia University Press.

Turner, L., & Ash, J. (1976). *The golden hordes: International tourism and the pleasure periphery.* New York: St. Martins' Press.

Williams, P., & Soutar, G.N. (2009). Value, satisfaction and behavioral intentions in an adventure tourism context. *Annals of Tourism Research, 36*(3), 415-438.

Photo: Jürgen Werner Kremer

Unconditioned Mind

What can be said
Of Unconditioned mind.
Some say that nothing can be said,
For in the saying,
Unconditioned mind is slain.

And yet, sometimes,
It comes,
It happens.

Sometimes appearing suddenly,
As though I were a jug of clay
Shattered by a stone.
At other times appearing slowly
As though the very substance of my being
Were like melting ice.
And again, once when I was dancing
It appeared as a remembrance of a lost land
Which seemed to be my true home.
Often in the deep forest,
It appears as though dawn had broken,
Suddenly, upon the darkest night.

Song of my heart,
Will you be able to sing
The song of unconditioned mind,
Sing that song
Which has no sound, no words?
Can you wait,
Song of my heart,
Through that eternity,
Until the soundless song is heard again?

Michael Sheffield